# CODE BREAKERS

# BREAKERS

# CODE BREAKERS

## THE SECRET INTELLIGENCE UNIT
that changed the course of the
## FIRST WORLD WAR

## JAMES WYLLIE AND MICHAEL MCKINLEY

EBURY
PRESS

1 3 5 7 9 10 8 6 4 2

Ebury Press, an imprint of Ebury Publishing
20 Vauxhall Bridge Road
London SW1V 2SA

Ebury Press is part of the Penguin Random House group of companies
whose addresses can be found at global.penguinrandomhouse.com

Penguin
Random House
UK

First published by Ebury Press in 2015

www.eburypublishing.co.uk

A CIP catalogue record for this book is available from the British Library

ISBN 9780091957735

Printed and bound in Great Britain by Clays Ltd, St Ives PLC

MIX
Paper from
responsible sources
FSC   www.fsc.org   FSC® C018179

Penguin Random House is committed to a sustainable future
for our business, our readers and our planet. This book is
made from Forest Stewardship Council® certified paper.

# CONTENTS

## PART III

# Introduction

# INFORMATION WAR

1916: The war that began in August 1914, and was meant to be over by Christmas, has become truly global. A famous author, now a British spy, sits at a street café in Spain telling a local police officer about the imminent arrival of a German submarine carrying biological weapons. On the mud-caked battlefield of the Somme, an artillery officer is training his guns on the Germans' own artillery, hoping that by smashing them he'll save the lives of thousands of his countrymen. In the hostile waters of the North Sea, an admiral guides the British fleet towards a titanic confrontation with the enemy's battleships. Surrounded by the seemingly limitless deserts of Mesopotamia, a general contemplates the best way to outmanoeuvre Turkish forces en route to Baghdad. Across the Atlantic, a tough New York cop is trying to uncover plots to destroy US factories. And as night descends on London, volunteers rush to man searchlights that will seek out Zeppelin raiders as they attempt to bomb the capital.

Each of these individuals was linked by a common thread: whether huddled together in nondescript rooms in Whitehall, hunkered down in makeshift accommodation close to the killing zones of the Western Front, or meeting in luxury hotels in Cairo, they were all acting on information supplied by codebreakers.

The codebreakers exerted a huge influence on the outcome of the First World War. Their achievements were the result of three interlocking factors: the genius of the codebreakers themselves; the emergence of the modern security state; and the recent revolution in communications technology – the wireless and the telegraph – that would usher in the first Age of Information.

They were an eccentric bunch, recruited mostly from academia – linguists and classicists were highly prized – but they also included writers, artists, theatre folk, crossword fanatics, retired military men and those who mixed in the same elevated social circles. Trained on the job, with no precedents to guide them, they were dedicated, tireless and determined to prove their worth, which they did time and time again.

As a result, Signals Intelligence – the interception and decoding of messages – established itself as the most potent branch of the emerging espionage agencies that had come into being in the years preceding the First World War.

The business of spying was still at a rudimentary stage in 1914. Over the next four years, the espionage community expanded hugely to try and cope with the demands of total war. Agent networks were established, tradecraft was perfected, and the collection of information was thoroughly compartmentalised. Human intelligence contributed to a number of very successful operations; however, overall, its impact on the war was negligible. If anything, the reputation of the burgeoning security services rested on the codebreakers' achievements and the intelligence supplied by them.

Of course, without material to work on, the codebreakers would have been irrelevant. New modes of communication, particularly the wireless, would provide them with hundreds and hundreds of messages. All the Great Powers were quick to take an interest in radio as a means of revolutionising their military communications. However, wireless had an Achilles heel: while messages could be transmitted and received, they could also be intercepted, the signals plucked out of the air. The advantages wireless offered were therefore offset by the fact that it was a fundamentally insecure medium for broadcasting sensitive information – and in war all information is sensitive.

This left the belligerents with no alternative but to encode their messages, which is where the codebreakers came in. Over the course of the war, all the combatants set up codebreaking organisations and they all achieved notable successes. The French were extremely effective on the Western Front; the

Germans cracked Allied naval and army codes; the Austro-Hungarians broke the Russian codes and vice versa; even the Americans, latecomers to the conflict, established their own 'Black Chamber' of cryptanalysts. However, none of them matched the sheer scale, scope and diversity of the British cryptographic effort. Overall, Britain's codebreakers achieved the most significant and wide-ranging results, ultimately changing the course of the war.

The British ran two codebreaking outfits: the navy had Room 40, so called because of the room number of their Admiralty home; while the army had Military Intelligence Section 1b – MI1(b) for short – based at the War Office. The two organisations had a truly global reach and dealt with thousands of enemy messages. They had a major impact on the war at sea, on land and in the air. They tracked the movements and actions of the German fleet in the North Sea and its U-boats as they prowled the Atlantic, while keeping a close eye on the Zeppelin threat. From the Western Front to the Balkans, North Africa to the Middle East, the codebreakers revealed the enemy's strategic and tactical plans, its strengths, weaknesses and whereabouts.

At the same time, Room 40 and MI1(b) exposed the secret machinations of the German government and its intelligence services as they plotted against Allied interests. The Germans targeted neutral countries across the world, especially America, using agents to foment sabotage and subversion. As a result of the codebreakers' labours, which included penetrating the diplomatic codes of over a dozen countries, Germany became more and more isolated and frustrated at its failure to hit the Allies where it hurt.

Without question, both Room 40 and MI1(b) made a decisive contribution to ultimate victory, their influence on the outcome of the conflict arguably greater than that of Bletchley Park during the Second World War.

Room 40, the naval codebreaking outfit, concentrated primarily on radio messages. MI1(b), the army equivalent, though it dealt with wireless more and more as the war progressed, handled mostly telegrams, the medium used by governments to communicate with each other. MI1(b) was gifted a huge advantage over its rivals because of Britain's dominance of international telegraph traffic. Determined to connect together its global empire, Britain had laid underwater cables across the seabeds of the world's oceans, with London acting as the hub of this communications network. As a result, nearly all telegraph messages passed through the capital on some stage of their journey, even those running through cables owned by other countries.

Though the Germans did have some international cable links of their own, they fell victim to some inspired forward planning by the British. In 1912, the Committee of Imperial Defence decided to cut all of Germany's underwater cables the minute war broke out between them. Near dawn on 5 August 1914, the *Alert*, a cable-laying ship, began dredging up from the sea floor all five cables that led from the German port of Emden to Vigo in Spain, then on to Tenerife and from there to the USA and the Azores. Within a matter of hours, the cables had been severed, and Germany had lost all telegraphic links to the world beyond continental Europe.

MI1(b) was also helped by the destruction of Germany's global wireless infrastructure. At Nauen, the Germans possessed an extremely powerful transmitter that could reach the wireless stations based in their African colonies, from where signals could be sent to the Americas. On 9 August 1914, the station and transmitter at Dar es Salaam was destroyed by the British navy; the Duala station in Cameroon was seized on 27 September, while the one at Windhoek in south-west Africa was put out of action by the Germans themselves.

Meanwhile, the rest of Germany's global chain was targeted. As Churchill, then the civilian head of the Admiralty, recalled: 'there existed in the Pacific only five German wireless stations, Yap, Apia, Nauru, Rabaul, Angaur, all of which were destroyed by us within two months of the outbreak of war'.

This brief but effective campaign meant that if the Germans wanted to communicate with the rest of the world, they had to use the telegraph. But with their own cables already cut, the only remaining option was to use ones belonging to neutral countries. Unfortunately for the Germans, the majority of messages travelling through these cables touched ground in London's telegraph offices at some point on their journey. Therefore, it was incredibly easy for MI1(b) to access this material; once requested, it was simply handed over. The many hundreds of messages gathered this way were then analysed by MI1(b)'s team of codebreakers.

While most of Britain was woefully unprepared for the war it was about to fight, a handful of individuals in Whitehall had the foresight to understand that the new modes of communication – the telegraph and the wireless – would have a significant impact on the outcome of any major conflict. However, the development of both MI1(b) and Room 40 into efficient and dynamic organisations was a much more haphazard affair, one that relied on the right people being in the right place at the right time.

# PART I

PART 1

# Chapter 1

# HEAD START

During September 1896, a couple of months before revealing his new creation – the wireless – to an astonished London audience, Guglielmo Marconi, the Italian inventor and entrepreneur, gave a trial demonstration to representatives of the British army and navy, sending a signal a mile and three quarters across Salisbury Plain. The men in uniform were suitably impressed.

Though the army flirted with using wireless, it was the Admiralty that embraced it enthusiastically. The benefits of being able to radio from ship to ship or ship to shore, regardless of how far they were from each other and their bases, were obvious. The British navy benefited hugely from an extremely close relationship with Marconi. Though not a truly original thinker, Marconi had a genius for absorbing the ideas of others and through inspired experimentation transforming them into something entirely new. He also had a keen business mind and an incredible flair for self-promotion. His international fame was secured in 1901, when, accompanied by a blaze of publicity, he sent a signal from the UK to Newfoundland; by 1907, he was operating a full transatlantic service.

Though keen to sell his product to as many nations as possible, including his native Italy, Marconi was a committed Anglophile. His loyalty to Britain was strengthened by his fear and loathing of Germany. He signed exclusive deals with the Admiralty, offered it access to his newest technological developments, including the interception apparatus that would grab German messages from the airwaves, and built a string of naval wireless stations along the British coast and across the Mediterranean. As a result, when war broke out, he was ready to give unconditional support, and his best scientists, engineers and technicians, to the British. His factories were taken over by the Admiralty, 5,000 of his wireless operators served the Royal Navy and the merchant fleet, his staff built and managed 13 long-range stations dotted across the world, and 348 Marconi personnel died in action, mostly at sea. Room 40 reaped the rewards of the Marconi Company's contribution to the war effort, gaining access to thousands of German wireless messages.

Sir Alfred Ewing, a distinguished scientist and inventor, was chosen to take charge of Room 40. A specialist in electrical engineering, Ewing became fascinated by earthquakes during a spell at Tokyo University, prompting him to create a new type of seismograph that measured tremors more accurately than previous models. In 1902, he was contacted by the Second Sea Lord of the Admiralty, Lord Fisher, who had embarked on an extensive and ambitious programme of reform and modernisation to meet the challenges of new technology. Fisher invited him to become the Director of Naval Education at Dartmouth College in Devon. Ewing observed that 'it is the educated man that the navy wants today, when appliances – appliances too, of an extremely complicated nature – are multiplying so rapidly in every department'.

*Marconi mounted motorcycle set*

His contribution was not forgotten, and on 4 August 1914 he was asked by the Admiralty, represented by Henry Oliver, then Director of Naval Intelligence (DNI), to set up a cryptographic unit. Much has been made of how these two old friends, meeting over lunch, casually got round to the subject of codebreaking, leading Oliver to make the off-the-cuff suggestion that Ewing might like to try his hand at running the show. Though the informal nature of this job offer fits in nicely with the amateurish, old-school-tie reputation of the British elite, the decision to approach Ewing was carefully calculated and showed an acute awareness of the fact that wireless had changed the nature of naval warfare.

Keen to do his bit, Ewing accepted the challenge. However, it quickly became clear that it was beyond his capabilities. Aged 59, and showing it, he slipped into the background of Room 40's day-to-day operations, a remote presence with little real control, and over time was slowly eased into retirement. Though his input was minimal, he played the part of necessary figurehead reasonably well.

Reginald 'Blinker' Hall, the man who would overshadow Ewing as leader of Room 40 and transform it into one of the most potent weapons in the Allies' arsenal, became the new head of Naval Intelligence (NID) in late 1914. Charismatic, energetic and cunning, called 'a genius' by the US ambassador, Hall would fully exploit Room 40's potential, and in doing so change the course of the war.

Hall had joined the navy in 1884, aged 14, and spent the next 30 years touring the globe and rising through the ranks. He got his first taste of espionage in 1909, on a training exercise in the Baltic, when he was ordered to spy on the harbour fortifications at Kiel, home of the German fleet. On arrival, he found the port seething with activity. It was regatta week and all the great and good were present, as well as the whole German navy. Writing years later, Hall recalled his initial dismay: 'I was not very confident of obtaining any results at all. It seemed impossible to get near the ships.'

But he was not a man to be easily discouraged. He asked the Duke of Westminster, who was there for the festivities, if he could borrow his high-speed motorboat, *Ursula*, for a couple of hours. The duke agreed, and next morning Hall took the boat out for a spin. Armed with a camera, he snapped away until he'd got all the information he wanted.

Though he clearly enjoyed it, thoughts of further intelligence work were soon forgotten once he was back on deck as commander of HMS *Natal*. In a service where those determined to resist change were as numerous and vocal as those trying to reform it, Hall was unafraid of technology and keen to utilise whatever benefits it offered. He also realised that showing concern for the welfare of his men was not an indulgence but a fundamental aspect of good leadership. One Room 40 veteran, William Clarke, remembered how he was 'most considerate to his staff, looking after their interests and their health in a way which endeared him to them and got out of all the best results'.

These qualities were apparent in 1913 when he took control of the *Queen Mary*, a new-style battle cruiser, extremely fast and heavily armoured. A deeply religious man, Hall broke with naval tradition and had a chapel built on board ship. Having attended to his men's spiritual needs, he improved their leisure time by providing a book stall and a cinematograph, one of the marvels of the age. No aspect of his men's lives escaped his attention. The regular sailors were expected to hand-wash their clothes themselves, while the petty officers, of whom there were nearly 100, paid to have their laundry done on shore. Hall considered this system a waste of time and money, so he contacted a company that manufactured washing machines, yet another new invention, and had one installed.

Not that he was a soft touch, far from it; a strict disciplinarian with hawkish features, he could be terrifying if he wanted to be. The nickname 'Blinker' came about because of a persistent twitch. What is striking in the testimony of those who knew him is how little this featured in their recollections. In others the twitch might have been taken as a sign of weakness and anxiety; not in Hall's case. Instead, what struck his colleagues about his piercing blue eyes was the effect they had when he trained his laser-like stare on you: once trapped by his unyielding gaze, it was impossible to escape the feeling that he was reading your mind. As Walter Page, US ambassador in

*Admiral Sir Reginald Hall, as commander*
*of HMS* Queen Mary, *1914*

London, observed, 'Hall can look through you and see the very muscular movements of your immortal soul while he is talking to you.'

As war approached, Hall was ready and eager to test the *Queen Mary* in battle. On 28 August, he was part of the squadron, under the command of the young firebrand Admiral David Beatty, which went to the aid of destroyers being harassed by enemy submarines and cruisers near the Heligoland Bight, the heavily mined exit point for the German fleet. During a brief engagement, two German ships were sunk.

This skirmish proved to be the last action he saw. Throughout his life he'd been plagued by a weak chest and bronchial problems that were aggravated by cold and damp weather. Given his current hunting ground was the North Sea, no stranger to freezing winds and driving rain, his condition soon became serious. On 10 September, Beatty noted that 'Captain Hall is far from well, he looks terribly grey and tired'.

Beatty was not the only one concerned for Hall's health. Ethel Agnes, Hall's wife, was so worried that she wrote to Admiral Oliver, who had just stood down as DNI to become Churchill's assistant, begging him to find her husband a desk job. Oliver obliged, and Hall took over as Director of Naval Intelligence. His frustration at being denied the chance to end his career at sea evaporated as he found himself at the helm of the good ship Room 40, and he proceeded to build an intelligence empire that spanned the globe.

In his unpublished autobiography, Hall explained his approach to the job, revealing a subtle grasp of human psychology and the need for flexibility: 'a Director of Intelligence who attempts to keep himself informed about every detail of the work being done cannot hope to succeed; but if he so arranges his organisation that he knows at once which of his colleagues he must go to for the information he requires, then he may expect good results. Such a system, moreover, has the inestimable advantage of bringing out the best in everyone working under it, for the Head will not suggest every move; he will welcome, and indeed, insist on ideas from his staff.'

The main criticism levelled at Hall by his colleagues was that he was too fond of intrigue. Without doubt, he relished devising elaborate ways to deceive the Germans and often made risky and hasty decisions. His secretary called him a gambler who enjoyed a dangerous game, and recognised the Machiavelli in him. Underneath, however, was nothing but a schoolboy; she fondly remembered how 'the fun and hazard of it all would fill him with infectious delight'.

Of all the intelligence chiefs, Hall exercised the most authority because he was able to escape the confines of his department. No door was left unopened, no area of policy neglected, no aspect of the war untouched. As a result, all roads led to him: Guy Gaunt, Hall's man in the USA, realised 'what a powerful friend he was … when I saw the men who came quietly into his office. I think I saw most of the cabinet, and for that matter, everybody else in England of any note.'

Hall's unique position owed a great deal to his forceful personality. However, it was the intelligence supplied by his codebreakers that justified his ubiquitous presence in the corridors of power. Credit goes to him for the way he marshalled this eclectic band of civilians, but it was their talent, ingenuity and perseverance that granted him the keys to the kingdom.

It was Churchill, as civilian head of the Admiralty, who laid the procedural and institutional basis for Room 40. Realising the immense advantages the codebreakers could provide, and equally aware that these would be rendered redundant if the Germans suspected that their communications were compromised, Churchill drew up a charter that guaranteed absolute security: the wireless intercepts were to 'be written in a locked book with their decodes, and all other copies are to be collected and burnt. All new messages are to be entered in the book, and the book is only to be handled under the direction of the Chief of Staff', a position occupied by Admiral Henry Oliver, a rigid workaholic who often did a 150 hour week; he was nicknamed 'the Dummy' because of his lack of facial expressions and monosyllabic utterances. Oliver would then circulate the decoded material to a handful of high-ranking naval personnel.

*The Old Admiralty Building, Whitehall, home to Room 40*

While this degree of caution was understandable, it was not particularly practical and it placed Room 40 in a straitjacket that was hard to struggle free from, delaying its growth into a fully fledged, proactive intelligence organisation. However, in its chaotic early days, this was the least of Room 40's problems as it stumbled into being.

Much of what we know about this period of Room 40's life comes from one of its first recruits, Alastair Denniston. Born in 1881, Alastair's father was a doctor who died of TB when he was ten. After performing well at school, he studied abroad, first at the Sorbonne in Paris and then at Bonn University, giving him a thorough knowledge of German, followed by a stint teaching languages at a school in Edinburgh and later at the Royal Naval College on the Isle of Wight: all ideal preparation for life in Room 40.

His pre-war claim to fame was as an athlete, playing in the Scottish hockey team at the 1908 London Olympics. The opening match took place

*Alastair Denniston, one of Room 40's earliest recruits*

on 29 October at Shepherd's Bush Stadium in atrocious conditions and pitted Denniston's Scotland against Germany. After winning the match 4–0, Scotland then faced England in the semi-final, played on the same day. The old enemy thrashed them 6–1 and went on to win gold by beating Ireland 8–1 in the final. A third-place play-off for the bronze medal between Scotland and Wales was abandoned after the Scots went home that evening; a historian of the sport concluded that 'the Scotland players had to get back to work'. In the end, a compromise was reached: both teams were awarded bronze and Denniston became the proud owner of an Olympic medal.

When Denniston joined Room 40, it only had five other staff. All of them, according to him, were 'singularly ignorant of cryptography'. Isolated from their boss, they had to carry their results down the corridor to his secretary. By November, this small group were installed in Room 40. Their new home was on the first floor of the Old Admiralty Building in Whitehall. Tucked away, quiet, with a view over an inner courtyard, its discreet location compensated for the cramped conditions.

A shift system was organised: two staff on duty between 9 a.m. and 7 p.m., then one on night watch, all engaged in the business of translation, sorting and decoding. Initially messages were brought by a 'never ending stream of postmen delivering bundles', then by pneumatic tubes (originally developed to speed up the exchange of telegrams between businesses), 'which discharged goods into a basket with a rush that shook the nerve of any unwitting visitor and much disturbed the slumbers of the night-watchman'.

Night duty in those first few months was a lonely time. There were no washing facilities and it was 'no good bringing pyjamas'. Instead, the night shift consoled themselves with 'plenty of sandwiches'.

Intercepted German wireless transmissions would provide Room 40 with the bulk of the material it analysed. This relatively new technology was available to the navy through its close contacts with the Marconi Company. Before August 1914, it already had wireless telegraphy (WT) interception masts at sites in Stockton, Chelmsford, Dover and London, plus access to

Marconi's stations, and had begun to pick up German communications; in the opening days of the war, a steady stream began to accumulate.

Considering the secrecy shrouding Room 40 – only a handful of top brass knew of its existence – it's hardly surprising that nobody thought to provide it with the messages that were coming in. Instead, it took the intervention of two well-heeled amateur radio enthusiasts to join the dots and connect Room 40 to its key source of intelligence.

Very few people in Edwardian England had the resources needed to own or construct a wireless set; those who did formed a socially exclusive club of hobbyists tinkering with primitive valves, amps and transmitters. One of these was Russell Clarke, a barrister, who began obtaining German intercepts. Quick to grasp their importance, he approached Sir Alfred Ewing, titular head of Room 40, and offered his services.

This was a fortunate chain of events, given the regulations laid out in the Defence of the Realm Act (DORA), which placed a plethora of restrictions on citizens' rights, including the right to own a radio. Police forces across the land were ordered to ensure that no domestic wireless appliances remained in operation. Quite how Clarke evaded these conditions is unclear; Denniston speculated 'that some rash official had tried his best on Russell Clarke and had been forced to retire worse for wear'. Clarke was soon joined by another radio ham, Bayntun Hippisley, who had also picked up signals 'obviously of Hun origin'. Clearly there was a rich seam to be mined, and interception apparatus was installed at Hunstanton coastguard station in Norfolk.

Clarke soon realised that 'he could intercept hundreds of … messages daily on short waves, which, if read, would give the daily doings of the German Fleet'. But Room 40 was still unable to make sense of them. Then the gods smiled on the codebreakers, not once but three times.

At 00.14 on 26 August, the *Magdeburg*, a German battle cruiser patrolling the Baltic, ran aground. Her captain ordered the destruction of two of her three copies of the *Signalbuch der Kaiserlichen Marine* (the Signal Book of the Imperial Navy), or SKM, leaving one available if they needed to signal for help. However, explosive charges that were meant to scuttle the vessel went off prematurely, and the crew began to abandon ship before the code books had been dealt with. At 04.10, the Russians showed up, and after three hours of inconclusive skirmishing, it was all over.

All three copies of the SKM were retrieved by the Russians, one from the ship and two from the sea. Exactly how this was done has never been conclusively established; Churchill, always ready to use poetic licence when the facts were lacking, wrote that 'the body of a drowned German ... was picked up by the Russians a few hours later, and clasped in his bosom by arms rigid with death, were the cipher and signal books of the German navy'.

The SKM was an immense tome containing 300,000 three-letter codes, each one referring to a plain-language word, phrase or name. Its covers were lined with lead, making it too heavy to hold in your hands. It had been in use for some time, pre-dating wireless, and included all forms of signalling employed by the fleet – flags, lamps and semaphore. Wireless codes, added later, were divided into sections featuring lists of ships, place names, compass bearings and map grid references.

The Russians understood the significance of the SKM. Knowing that the British navy would profit handsomely from it, they entrusted a copy to the son of the Russian ambassador in London, who gave it to the Admiralty on 13 October.

The next stroke of luck came courtesy of the Royal Australian Navy. On 11 August, they interned a German merchant ship in Melbourne harbour. On it was a copy of the *Handelsschiffs Verkehrsbuch* (HVB), a signal book that used four-letter codes. A month later, this copy was forwarded to the UK, arriving at the Admiralty in late October. Denniston recalled how Room 40 soon discovered that the HVB was used by the whole German fleet, including its submarines and airships.

The final find occurred during November, when British trawlers working in the neighbourhood of the spot where four German destroyers had been sunk close to the Dutch island of Texel, recovered a code book used by the German Admiralty and senior officers. The *Verkehrsbuch* (VB), which contained 100,000 code groups, each consisting of five numbers, was used for correspondence with the naval attachés abroad, especially in Madrid, where the German embassy was orchestrating sabotage and subversion across the Mediterranean and North Africa.

Thanks to this 'magnificent draft of fishes', as the haul was dubbed, Room 40 staff now had the means to decode everything the Germans said. They were helped immensely by the German reaction to these potential security breaches. On the day after the sinking of the *Magdeburg*, its squadron commander admitted that it was not known whether the SKM had been destroyed. However, an official hearing convened in September concluded that there was no evidence that the Russians had got hold of it. Prince Henry of Prussia, commander of the Baltic fleet, wasn't so sure. He thought it was 'probable' that the Russians had recovered one of the signal books. Nevertheless – and incredibly – the SKM was not replaced until May 1917.

Not long after the Australians took possession of the HVB, the governor of German South-West Africa and various cruisers roaming the South Atlantic were informed that the book was compromised. Despite this warning, nothing further was done and the HVB stayed in use until 1916. As far as the VB was concerned, it never occurred to anyone that it might have fallen into the lap of the enemy, and it too was kept in operation until 1917.

The Germans' inadequate and lethargic response to these incidents, coupled with the arrogant assumption that their codes were infallible, cost them dear. Room 40 had ample time to thoroughly familiarise itself with the habits, movements and intentions of the enemy's battleships, U-boats and Zeppelins, while their networks of spies, conspirators and terrorists were similarly exposed. By the time the books were finally replaced, the British codebreakers had become so adept and experienced that they were able to tackle them head on.

## Chapter 2

# BATTLE STATIONS

At around 8 a.m. on 16 December 1914, two ships from a squadron of German battle cruisers under the command of Admiral Franz von Hipper opened fire on the Yorkshire resort of Scarborough. At the Grand Hotel overlooking the sea, one of the guests was shocked awake by 'a tremendous noise' and 'looked out of the window and saw a huge flame and clouds of smoke'. As shells crashed into the town, thousands fled their homes and headed for the railway station. Half an hour later, the bombardment was over, leaving 12 dead – including a postman blasted to smithereens while delivering mail – 99 wounded, and churches, public buildings and homes reduced to smouldering ruins.

At roughly the same time, Hartlepool, an industrial town with a busy dockyard, 65 miles to the north of Scarborough, also came under fire from German ships. More than 300 houses were damaged, with 86 civilians killed and 424 wounded. An hour later, Whitby, 20 miles north of Scarborough, suffered a 10-minute bombardment and the loss of two lives.

Forty-eight hours earlier, Room 40 had picked up indications that the Germans were up to something. For the very first time, the codebreakers would have a major influence on the planning and tactics of a naval operation. According to First Lord of the Admiralty Winston Churchill, decoded messages suggested 'an impending movement which would involve battle cruisers, and perhaps … have an offensive character against our coasts'.

At this point, Room 40 had received no indication that the bulk of the German High Seas Fleet, under the command of Admiral Friedrich von Ingenhol, was also going to be involved in the operation. As a result, only sufficient forces to deal with Hipper's battle cruisers were put out to sea. This was a mistake. The German High Seas Fleet was indeed following in the wake of Hipper; it was to play a supporting role, either by offering assistance or by pouncing on any British ships that strayed across its path. However, a few hours before Hipper reached his targets, Ingenhol received information that the British were also on the move, and promptly turned his ships round and headed back to base.

Nevertheless, the codebreakers had still created an opportunity to catch Hipper unawares. To retain the element of surprise, no attempt was made to prevent the raids from being carried out. After all, if the British were ready and waiting, the Germans might suspect that their codes had been broken. Instead, the British cruiser force was to lie in wait for Hipper as he made his return journey. Due to poor weather and the confusion it caused, however, they were unable to engage him and he slipped safely back to Germany.

Though the Admiralty was disappointed that Hipper had escaped unscathed, and deeply embarrassed by the public's justifiable outrage at its failure to defend the coast, Room 40 had proved its worth. Churchill drew comfort from the fact that 'the indications upon which we had acted had been confirmed by events. The source of information upon which we relied was evidently trustworthy.'

The challenge facing Room 40 and the Royal Navy in 1914 was shaped by one of the great strategic military follies of the pre-war period, or indeed any period: the Kaiser's quest to build a fleet to rival the Royal Navy. A lover of all things nautical, he jealously eyed the mighty British fleet and wanted one of his own to play with. Exactly what purpose his navy would serve, beyond satisfying his childish impulses, was not clear. More than anything else it convinced the British that the Germans were out to get them, and they responded in kind with their own massive building programme, which included constructing the biggest, most powerful warships the world had ever seen, the dreadnoughts, whose weight of fire was sufficient to wipe out any other vessel in existence thereby forcing the Germans to do the same.

Despite Germany's huge commitment of money and manpower, it was not enough to achieve parity with the Royal Navy, and as the war began, Britain retained a decisive overall advantage, able to deploy 121 cruisers and 221 destroyers against Germany's 40 cruisers and 90 destroyers. Crucially, in the North Sea, the British had 20 dreadnoughts to Germany's 13. Closing that gap was nigh impossible: the British had more ships under construction and would continue to produce more than the Germans as the war wore on. The British ships were also faster and armed with heavier guns.

Many expected the British to capitalise on their superiority by seeking an immediate confrontation with the Germans. Lord Jellicoe, commander-in-chief of the British Grand Fleet, was not so gung-ho. As a veteran admiral with considerable experience of the German navy, he was fully aware of its strengths and had no intention of underestimating his enemy. His strategic thinking was dictated by a realistic assessment of the threat the High Seas Fleet posed, combined with his fear of the sudden and unsuspected destruction that the new weapons of naval war, the torpedo and the mine, could cause: on the first day of the war alone, the Germans laid 25,000 mines in the North Sea.

For Jellicoe, the preservation of British naval supremacy was paramount. The simplest way to safeguard it was to remain on the defensive and venture out only when necessary; the mere presence of the fleet enough to deter the Germans and maintain the maritime blockade that Jellicoe believed would ultimately win the war.

Positioned at the entrance to the North Sea and the Channel exits, the British navy exerted a stranglehold on Germany's overseas trade, cutting it off from the rest of the world. Any ship carrying goods to Germany could be stopped and taken out of commission. By the middle of 1915, nearly half the total tonnage of Germany's merchant fleet was marooned in neutral ports, unable to leave. Given Germany's reliance on imports, this was a serious blow: 25 per cent of its dairy, fish and meat came from abroad; its potato crop depended on fertilisers, 50 per cent of which were imported from the USA, Chile and North Africa; while the Americans provided nearly all of its cotton, large quantities of wheat and 60 per cent of its copper.

*Admiral Jellicoe, commander of the Fleet during
Jutland and the U-boat campaign*

The Kaiser, reluctant to see his toys get broken, and hoping that his army would deliver a speedy victory, thereby rendering the blockade null and void, initially decided that the High Seas Fleet should remain as close to its bases as possible and not seek battle with the British. This was hard to swallow for many in the German navy, as they knew they were superior in a number of vital areas. Their ships had far thicker protective armour and as a result were less easy to sink; their gunnery was more accurate – in all confrontations with the British, they scored more hits on target; their optical equipment and rangefinders were more advanced and their searchlights more sophisticated, giving them greater capability to fight at night.

But as it became clear that the ground war would continue into 1915, the German naval high command managed to convince the still cautious Kaiser to let them off the leash a little. The aim was to nibble away at the

British Grand Fleet by luring portions of it out to sea and into the arms of the German High Seas Fleet. Over time, operations like Hipper's assault on Scarborough, the first of its kind, might reduce the deficit between the navies and provide the Germans with the opportunity to engage the enemy in a fair fight.

The German approach relied on surprise and deception. However, if the codebreakers in Room 40 could supply accurate predictions of the movements of the High Seas Fleet, they would not only deprive it of any tactical advantage but might also allow the British to engineer a situation that allowed them to catch a section of the German fleet away from its bases and expose it to the full force of Jellicoe's dreadnoughts.

Not that the Germans lacked wireless interception and decoding capabilities of their own. In late 1914, after army units had picked up and decoded British messages emanating from Dover, the navy decided to get in on the act, establishing their main wireless interception station at Neumünster and forming a deciphering bureau, the Entzifferungsdienst, or E-Dienst.

At first this was staffed by naval men with no cryptographic experience or talent; only later did they draft in academics. Though the E-Dienst achieved some success, the British naval codes were generally more secure and changed more frequently than the German ones, and the Grand Fleet maintained better wireless discipline while at sea, running silent for most of the time. Critically, the Germans never considered the possibility, despite mounting evidence to the contrary, that their own codes had been broken.

As the number of interception aerials multiplied, the sheer volume of transcriptions of wireless messages, sent directly from the listening stations, and arriving by tube at Room 40, was becoming hard to handle. Three copies were made of messages that were processed successfully. Those that weren't – a 'vast number of fragments ... in unknown codes or languages' – were collected in a large tin, which was labelled 'NSL: Not Logged or Sent'. In the minds of the codebreakers, the tin took on a life of its own, invading the dreams of one nightwatchman, who woke trembling in a sweat after a nightmare in which he had been put in the NSL and got lost. A constant reminder of what Room 40 was failing to accomplish, the tin soon became an object of hatred.

This mountain of stuff increased dramatically with the advent of direction-finding stations. Marconi had begun experiments on DF apparatus, which could pinpoint the origin and location of a wireless transmission, as early as 1905, and the results had been tested on a Cunard liner in 1912, but the system wasn't perfected until 1914.

The man responsible for bringing DF technology to fruition was Captain Henry Joseph Round, one of Marconi's key men and a true radio pioneer. With a first-class degree from the Royal College of Science, he joined Marconi in 1902 and achieved breakthroughs in the design of vacuum tubes, cathodes and diodes. By the end of his career he had acquired 117 patents; his work during the war on valve amplifiers vastly improved interception capability.

Seconded to the army, Round introduced DF stations to the Western Front. Then, in February 1915, he visited Hall at the Admiralty. Excited by the benefits DF could deliver, enabling Room 40 to pinpoint the location of battleships, U-boats and Zeppelins from the wireless signals they sent, Hall ordered Round to erect a station for the Admiralty. A site was chosen at Lowestoft, in Suffolk, and before long the number of DF facilities grew to six, dotted along the UK coastline.

By now, Room 40 desperately needed more staff. Recruitment was Hall's domain, and as he acknowledged years later, when drafting his autobiography, the best candidates were 'men who in normal times would have laughed at the idea that they could ever be of conceivable use to our Intelligence Services'. Hall utilised his social connections and unearthed writers, an expert on furniture and art, and several language teachers.

His main focus, however, was on academia. Confronted with the problem of having to decide what area of expertise might produce the best codebreakers, Hall concentrated on those with a background in alien languages, modern and ancient, on the basis that they would be familiar with the challenge of untangling the meaning of texts, grappling with unusual terms and syntax, and employing the sort of analytical rigour that codebreaking required.

One of the new recruits with a linguistic background was Nigel de Grey. Educated at Eton, the son of a rector, and fluent in French and German, de Grey had been working for the publisher William Heinemann before the war. Twenty-eight in 1914, he served in the Royal Naval Volunteer Reserve (RNVR) in Belgium during a doomed attempt to stop the Germans capturing Antwerp. The operation, Churchill's brainchild, came too late to save the city, and the remnants of the small British force were left to fend for themselves, grabbing any opportunity they could to escape.

Back in the UK, de Grey was head-hunted by Hall and became one of Room 40's most important codebreakers. His chief collaborator was Reverend William Montgomery, an expert in medieval German theology from Cambridge who had translated several obscure religious texts from that period into English.

Hall also scoured the universities for suitable individuals with classical training: men like Frank Adcock, future Professor of Ancient History at Cambridge, and Alfred Dillwyn Knox, better known as Dilly. By the general agreement of his Room 40 contemporaries, and those who continued their trailblazing work, Dilly Knox was the most gifted codebreaker of them all.

Skinny, bespectacled, and original in everything he did, Dilly perfected a style of spin bowling that bamboozled countless cricketers. He produced similar confusion when playing bridge, bewildering his opponents with his choice of cards. Riding his beloved motorbike, he was a danger to himself and others. In 1931, after years defying the odds, he had a serious crash and broke his leg. After that he got a car, an Austin 7, which he drove to High Wycombe station every day to catch the London train. During the 15-minute journey, he would recite Milton, often with his hands off the wheel. Almost nobody wanted a lift from him; one brave resident recalled how Dilly 'used to amuse himself seeing how far he could go downhill with the engine off'.

Born in 1884, Dilly came from an ecclesiastical family. His grandfather was a wandering missionary who spent many years in India preaching to the poor and destitute; his father was an Anglican priest, and his mother a Catholic of Norman descent. Two of his three brothers would serve Rome, one as a populist writer and commentator on spiritual and moral issues, the other ministering to the dispossessed of east London; his other brother settled for secular pursuits, becoming a much-admired editor of *Punch*.

Growing up, the boys enjoyed inventing games and puzzles, and were obsessed with trains, learning the timetables off by heart and testing each other on them. It's tempting to think that Dilly's uncanny ability to hold vast amounts of abstract data in his head and imagine it arranged and rearranged in rows and columns came from studying *Bradshaw's Railway Guide*.

Dilly was seven when the family exchanged the rural parish of Knebworth for the grim, dirty industrial district of Aston in Birmingham, crammed with factories and squalid terraced housing. His father threw himself into his work, and the presence of a railway track at the bottom of their garden delighted the boys, as did the thrill of riding trams round the city. However, during their first Christmas there, his mother came down with flu, and eight months later she was dead. Stunned by this terrible blow, his father packed Dilly off to stay with a widowed great-aunt in Eastbourne.

There he attended a local prep school, then Summer Fields near Oxford, before earning a scholarship to Eton, where he excelled in mathematics and classics. During the entrance exams for King's College, Cambridge, he submitted two brilliant papers in Greek and maths, but didn't bother to complete the others. Nevertheless, they offered him a place.

By now a pipe-smoking atheist with an interest in amateur dramatics, Dilly came under the influence of Walter Headlam, a brilliant but eccentric classics scholar who never took the right train, once rode a horse into a pond, and only used stamps that he liked the colour of. Headlam's great project was deciphering, reconstructing and translating Herodas' *Mimes*, a series of often obscene satirical dialogues that had been discovered during an archaeological dig in Egypt in 1889 and deposited in the British Museum. This selection of worm-eaten, faded and incomplete papyri had been transcribed by a slave around AD 100, which only added to the difficulties facing scholars who attempted to make sense of them.

Undaunted, Headlam was determined to produce a satisfactory edition and enlisted Dilly's help.

In June 1908, however, Headlam, always a frail creature, died suddenly at the age of 42. Dilly, having been made a fellow of King's College in 1909, continued his mentor's work, travelling back and forth from Cambridge to the British Museum. The job was all-consuming. He had to get the fragments of text in the right order – not easy given that there were chunks missing – fill the gaps where words and letters were absent, use inspired guesswork to account for the copyist's mistakes, and, as there were no breaks in the script, decide which characters were speaking at any given point in the dialogues.

Despite these hurdles, Dilly made good progress until the war interrupted his efforts and he joined Room 40. As it turned out, the painstaking process of deciphering Herodas was the perfect training for life as a codebreaker. Almost immediately he demonstrated his instinctive flair for textual analysis by bringing to bear the methodology he'd practised when interpreting ancient Greek poetry. He set about scrutinising less sophisticated German communications, such as weather reports, for any careless errors made by the wireless operators during transmission – inaccurate spellings, short cuts or sloppy habits – that might provide clues to the cipher keys being used in more complex messages. This technique proved effective, and Dilly continued to use it throughout the war, with considerable success.

After the German navy's initial attempt to catch the British Grand Fleet off guard during their attack on Scarborough, the next foray by Hipper's battle cruisers occurred in late January 1915. On the 23rd, Room 40 got wind of it and Churchill was told that 'these fellows are coming out again … and a raid upon the British coast was clearly to be expected'.

In fact Hipper intended to lurk near Dogger Bank with the rest of the German High Seas Fleet close by, and wait for the British battle cruisers to arrive, which they duly did. Unfortunately for Hipper, Ingenhol, his commander-in-chief, did his disappearing act again and withdrew, leaving him outnumbered by Beatty's ships. During an inconclusive engagement,

Hipper lost the *Blücher* but managed to escape further punishment when Beatty, wrongly believing that U-boats were in the vicinity, did not pursue his prey.

Despite once again proving its worth, Room 40 was generally regarded with suspicion by naval personnel. What could a bunch of civilians possibly know about the complexities of war at sea? Clearly, during the first few months of the war, this was a valid concern; as Denniston freely admitted, 'of naval German, of the habits of war vessels of any nationality, they knew not a jot'. Schoolboy errors were made that seemed to confirm the prejudices of the Operations Division, whose job it was to interpret Room 40 material. On one occasion, Room 40 reported on the movements of a German ship, the *Ariadne*, which had in fact been sunk a few weeks earlier.

Incidents like this gave the Operations Division every excuse to exclude Room 40 from decision-making. When the codebreakers asked for a flagged map of the German coastline, the Operations Division refused. As Denniston caustically observed, 'no attempt was made to develop the intelligence side of the work'.

Blinker Hall was the first to recognise the problem and take steps to address it. On 16 November 1914, he appointed a naval expert, Captain H. W. W. Hope, to act as Room 40's guide and improve their intelligence analysis. With the volume of intercepted messages increasing and the direction-finding network expanding rapidly, Hope was able to put his knowledge to good use. Luckily, the Germans made life easy for him, as he was happy to acknowledge: 'experience showed that the Germans were exceedingly methodical in their methods and a large number of routine signals were made day after day, which were of great assistance'.

This accumulation of data gave Hope and his team 'a good working knowledge of the organisation, operations and internal economy of the German fleet', and he was proud of the fact that nothing out of the ordinary happened without some kind of warning. Though Room 40 had begun as a skeleton crew struggling to come to terms with the unique language of naval warfare, it had in a short time become a well-oiled machine with an intimate knowledge of the enemy.

However, this transformation was not reflected in the processing of its intelligence. Decodes were still handed over with minimum comment to the Operations Division; there they were scrutinised by Henry Oliver, Chief of Staff, who personally decided what to show the Admiralty top brass, who then agreed amongst themselves what to tell their boss, Admiral Jellicoe.

Frustrated by a system that often meant he received important decodes too late or not at all, Jellicoe asked for the material to be sent directly to him. His request was denied. The negative consequences of sticking to a procedure set up in the first few weeks of the war, when Room 40's potential was an unknown quantity, would be fully felt in 1916 at the Battle of Jutland, one of the largest naval actions in history.

But for now, all attention was focused on the German U-boat campaign under way in the Atlantic, where the Kaiser's submarines were sinking large quantities of British shipping and threatening to sever the vital trade link with America.

Chapter 3

# DEATH ON THE ATLANTIC: THE LUSITANIA

The potential impact of Germany's U-boats was severely compromised by the rules of maritime engagement. Under international agreements, submarines approaching a civilian vessel were required to announce their presence and confirm without a shadow of doubt what the ship was carrying before engaging it. This procedural limitation blunted the U-boats' offensive capability.

In early 1915, after much heart-searching and head-scratching, the Germans took the fateful decision to ignore these conditions and embark on unrestricted submarine warfare against vessels sailing in waters around the British Isles (except for a route north of Scotland): no warnings, no special pleading; any boat was fair game if the U-boat commanders suspected its cargo included war material, whether it be food, raw materials or weapons. The discretion to sink a ship, neutral or otherwise, was theirs.

It was a gamble, and a dangerous one. The chances of avoiding collateral damage while prosecuting an unrestricted submarine campaign vigorously enough to attain Germany's goal of forcing Britain out of the war by crippling its trade with America were slim. The risk that civilians would become casualties was high. The unknown quantity was how America would react if German U-boats attacked American interests. As far as the Germans were concerned, it was a risk they had to take, for ships leaving neutral American ports were bringing war supplies to Germany's enemies, so the submarine warfare policy, Germany hoped, would act as a deterrent to the USA, and not a provocation.

The man charged with performing this balancing act was the German ambassador to the United States, 51-year-old Count Johann von Bernstorff, a tall, polished, cold-eyed charmer with a blonde moustache. One of his first actions after the war began was to hold court for journalists in his suite at the Ritz-Carlton, a three-year-old luxury hotel in midtown Manhattan, conveniently located near all the New York social clubs of which von Bernstorff was so fond, as well as a short taxi ride away from the German Club on Central Park South. As newspapermen peppered the count with questions about the war in Europe, he paced the room excitedly, explaining to them in his flawless, witty English that they and their journals were wrong about the ferocious and unprecedentedly fatal battles being fought in Europe: the French were thoroughly beaten, the German invasion of Paris was imminent, and in any case, the Russians had started all the trouble.

*Count Johann von Bernstorff, the German ambassador*
*to the USA, 1908–17*

This was the first shot in a publicity war that von Bernstorff would wage as he tried to construct the 'right kind of news' for America by planting stories in pro-German newspapers like *The Fatherland* and *Staats-Zeitung*, throwing money at the *New York Evening Mail* and even attempting to buy the *Washington Post*.

Von Bernstorff had been recalled to Berlin in early July, shortly after the assassination of heir to the Austro-Hungarian throne Archduke Franz Ferdinand and his wife, the flame that lit the fuse of war. While there, he met with Lieutenant Colonel Walter Nicolai, head of *Abteilung Drei-Bai*, or Section 3B, Germany's military intelligence unit. Nicolai explained to the ambassador that all of the country's military intelligence officers had been deployed to the conflict zones of Europe. It was now up to von Bernstorff to add to his diplomatic duties the most undiplomatic of missions: he was to become Germany's unofficial spymaster and saboteur in North America. To help him on his way, von Bernstorff received $150 million in German treasury notes (worth almost $3.5 billion today), in order to buy 'munitions for Germany, stopping munitions for the Allies, necessary propaganda, forwarding reservists – and other things'.

Von Bernstorff's chances of being able to keep the United States out of the war were boosted by US President Woodrow Wilson's isolationist attitude to the European conflict. On 4 August 1914, the day that Britain declared war on Germany, Wilson, who was under heavy stress due to his beloved wife Ellen's terminal nephritis, which would kill her two days later, issued a detailed proclamation of American neutrality, which barred any American or anyone living in the United States from aiding any of the belligerents in prosecuting the war. Even so, in reality, the USA's economic machinery soon started to work in favour of the Allies.

When war broke out, the United States was suffering from an industrial recession and a bear market. The intense lobbying of J. P. Morgan & Company and other financiers in Washington and London would by October see the US allow sales of war products to the combatants, which of course meant the Allies, as the Germans couldn't get through the British blockade. On top of that, the British awarded all their US war purchasing to J. P. Morgan, to prevent war profiteering and create price stability, a contract that saw Morgan take a staggering commission of $30 million. The relaxing of restrictions also created a boom for American manufacturers, farmers and banks, and would see the US replace England as the world's financial superpower by the war's end.

As far as many Germans were concerned, these developments effectively made the US a belligerent in the conflict: an enemy power and therefore a legitimate target, not just at sea but also on home soil. At the same time, the countries on its immediate borders, Canada and Mexico, offered opportunities for causing the Allies trouble.

Canada, to the north, was a young, muscular and resource-rich Dominion of Great Britain eager to assert its own identity in the war with Germany; Mexico was aflame with a revolution that had begun in 1910 and that would last for a decade, illuminating its long, bloody history with the US. And the United States itself, with nearly 100 million people, of whom 2.3 million were German-born immigrants, was a wealthy, diverse and powerful base from which to work German propaganda – or outright sabotage against America's allegedly neutral war machine.

Von Bernstorff had spent the first 11 years of his life at the Court of St James in London, where his father, Count Albrecht, served as Germany's ambassador to the United Kingdom. When his father died in 1873, von Bernstorff moved back to Germany, and graduated with a baccalaureate from Dresden before joining the diplomatic service. He took the fast track: from military attaché in Constantinople in 1899 to first secretary in London in 1902, where he caught the eye of Kaiser Wilhelm as a man who could win goodwill for Germany with his keen political skills and considerable personal charm. After serving as consul general in Cairo from 1906–8 – a launch pad to the diplomatic big league – von Bernstorff was appointed Imperial German ambassador to Mexico and the United States in 1908. Along the way, like so many other European aristocrats who had ancient titles to dangle, he acquired a wealthy American wife (and eventually an American mistress).

The first real test of his capacity to calm American fears about German intentions would be brought about by U-boat action in the Atlantic. On

OCEAN STEAMSHIPS.

# CUNARD

### EUROPE VIA LIVERPOOL
# LUSITANIA
Fastest and Largest Steamer
now in Atlantic Service Sails
SATURDAY, MAY 1, 10 A. M.
Transylvania, Fri., May 7, 5 P.M.
Orduna, - - Tues., May 18, 10 A.M.
Tuscania, - - Fri., May 21, 5 P.M.
LUSITANIA, Sat., May 29, 10 A.M.
Transylvania, Fri., June 4, 5 P.M.

Gibraltar—Genoa—Naples—Piraeus
S.S. Carpathia, Thur., May 13, Noon

## NOTICE!

TRAVELLERS intending to
embark on the Atlantic voyage
are reminded that a state of
war exists between Germany
and her allies and Great Britain
and her allies; that the zone of
war includes the waters adja-
cent to the British Isles; that,
in accordance with formal notice given by the Imperial German Government, vessels flying
the flag of Great Britain, or of
any of her allies, are liable to
destruction in those waters and
that travellers sailing in the
war zone on ships of Great
Britain or her allies do so at
their own risk.

**IMPERIAL GERMAN EMBASSY**

WASHINGTON, D. C., APRIL 22, 1915.

*The German warning to passengers sailing on
transatlantic ships of the dangers they faced*

1 May 1915, the day that the *Lusitania* set sail on her final journey, 40 newspapers across the United States published a chilling notice from the Imperial German embassy in Washington DC, one echoing those it had published since launching their submarine war on Atlantic shipping. Its message was lethally simple: 'travellers sailing in the war zone on ships of Great Britain or her Allies do so at their own risk'.

The *Washington Times* splashed the warning at the top of their front page, next to a photograph of a sinisterly arrogant von Bernstorff, and an ominous piece reported how nine prominent passengers about to set sail on the Cunard liner the RMS *Lusitania* from New York to Liverpool had received anonymous telegrams threatening trouble on the seas.

'Alfred G. Vanderbilt was told in one of these messages that the vessel would be torpedoed,' the *Times* article revealed. 'Other passengers were warned that the liner would meet some mysterious end. The messages were 'followed up' by the circulation, by a number of strangers on the crowded pier, of similar veiled warnings. The strangers hurried away as soon as the fact that they were accosting passengers was reported to the Cunard private detective force.'

Vanderbilt had inherited the patriarchal place at the Vanderbilt family table when his father, Cornelius II, died in 1899. He had cancelled his booking on the *Titanic* three years earlier at such a late hour that some reports of its destruction listed him as one of the dead. Now, however, he and other well-heeled travellers were being warned by mysterious forces: 'THE LUSITANIA IS DOOMED. DO NOT SAIL ON HER,' read the telegram received by Vanderbilt, its macabre signature, 'MORTE'. Death.

The *Lusitania* had been a German target since the beginning of the war. The British Admiralty had registered the ship as an armed merchant cruiser in September 1914, planning to fit guns on her decks. The plan was scrapped as too energy-expensive, but the idea created the abiding impression that the liner was not a luxurious vessel for civilians but a weapon of war. The fact that the *Lusitania* was a target was not news to the codebreakers in Room 40. In March 1915, Blinker Hall's team had been reading intercepted German intelligence reports detailing ships heading for British ports, as well as those heading to the neutral Netherlands. On 2 April, after the Germans had sunk neutral Dutch, Spanish, Norwegian and Greek ships, much to their governments' outrage, the Kaiser declared that the ships of neutral countries would not be targets. On 10 March, Room 40 intercepted German intelligence broadcast from the high-power long-wave station in the North Sea town of Norddeich reporting that the 'fast steamer *Lusitania* leaves Liverpool March 13th'.

While the *Lusitania* was clearly an enemy ship insofar as she flew Britain's nautical flag the Red Ensign, the ominous telegrams sent to prominent passengers in New York imply that her targeting had more sinister motives, a suggestion that has led to many conspiracy theories since her doom.

One of the most compelling is that Room 40, while not having a direct hand in her destruction, certainly helped create a climate of terror around her. The telegrams sent to Vanderbilt and others all originated at the *Providence Journal*. This small local newspaper punched far above its weight during the First World War, often breaking stories then picked up by global powerhouses such the *New York Times*. The *Providence Journal*

*Captain Guy Gaunt, the British Naval Attaché to the USA*

was run by John Revelstoke Rathom, who had risen from being the paper's managing editor, to running the operation as editor and general manager in 1912. Rathom was an Australian ex-pat who happened to be great mates with another Australian ex-pat, Guy Gaunt, the British Naval Attaché in Washington DC.

Guy Gaunt, like other foreign service operatives, spent much of his time in New York City, which fuelled both his access to and his love of society, and his considerable ego. While not trained in intelligence, he was Britain's de facto spy for the entire United States for the first two years of the war, reporting back to Blinker Hall first, and MI5, second. As such, Gaunt would have access to Room 40's intelligence and through his friendship with Rathom, a friendly newspaper in which to plant whatever seeds he needed to sow.

The threatening telegrams arrived on the same day as the warning issued by the German embassy, which was likely penned for von Bernstorff by his unofficial propagandist George Sylvester Viereck, editor of the pro-German five-cent weekly *The Fatherland*. On the cover of its 28 April 1915 edition it featured a drawing of an Allied liner not unlike the *Lusitania* – though with a gun prominent on her bow – listing after being hit by a torpedo. The caption beneath the depiction of survivors in lifeboats and floundering in the water read: 'The Work of a German Submarine'. And, with no small irony, beneath that the title of the edition's featured article was 'German Love of Peace'.

Those who boarded the *Lusitania* on 1 May 1915 might have taken comfort in a *New York Times* article accompanying von Bernstorff's warning, in which Cunard's agent, Charles B. Sumner, claimed that when he first heard the warning submitted over the telephone the previous night, he thought it was another blackmailer. Sumner reported that Cunard had received several blackmail attempts against its Atlantic liners, most recently one demanding $15,000 to prevent a similar notice threatening harm, which he dismissed as a nuisance 'to annoy the line and make its passengers uncomfortable'.

Sumner had no fear for the safety of Cunard's passengers, and indeed stressed the security measures that the line employed.

'No passenger is permitted aboard … unless he can identify himself. No express matter (i.e. unaccompanied parcel) of any sort is taken. Every passenger must identify his baggage before it is placed on board.' He added that the British navy was responsible for all ships in the danger zone off the British Isles, and 'especially for Cunarders … As for submarines I have no fear of them whatever.'

Anyone sceptical about a Cunard employee practising damage control would have taken no solace in another *New York Times* piece that day, which reported that a German submarine crew had mocked the survivors of a ship they had torpedoed. B. T. Peak, the second engineer of the British steamer SS *Falaba*, which had been sunk by submarine U-28 off the Irish coast on 28 March – resulting in 100 dead, including an American – said from his London hospital bed: 'I was hoping they would pick me up but

The Fatherland *28 April 1915*

41

instead they were laughing and seemed to treat it all as a huge joke … It was quite evident the Germans were prepared to see the people drown.'

The Germans had sunk 19 merchant ships in the Atlantic between October 1914 and the beginning of unrestricted submarine warfare in February 1915, and that total was about to rise sharply (by the end of the war they would have sunk 5,000 Allied ships). The British Admiralty sent destroyers to escort passenger ships, and instructed them not to fly flags or otherwise call attention to their nationality. The *Lusitania* had her orange and black funnels painted dark grey during the voyage to make her well-known profile less visible to submarines, but Cunard had also shut down one of her four engine boiler rooms to save money on light passenger loads, reducing the ship's top speed by three and a half knots, to 22 knots, or about 25 miles per hour.

Nearly one week on from the sinking of the *Falaba*, after the first-class passengers had taken their Friday lunch in the frescoed gold and white Louis XVI dining room, the *Lusitania* was on the home stretch of her journey, sailing in patchy fog about 11 miles off the fishing village of Kinsale, on the south coast of Ireland. The liner's captain, William Turner, who had spent 46 of his 59 years at sea, had reduced speed, partly due to the fog, and partly to time his arrival in Liverpool for after dark, when it would be harder for a submarine to target the ship. Despite Sumner's assurance, there were no British cruisers escorting the *Lusitania* safely through the danger zone.

The German submarine U-20, however, *was* sailing in those waters, hunting for Allied shipping. The codebreakers of Room 40 knew the vessel was active, and so too did the Admiralty. The U-boats maintained regular wireless contact with their home bases, messages that were intercepted and then decoded by Room 40. This accumulated data gave the codebreakers intimate insights into the habits and patterns of the U-boats – when they left port, their direction and speed, until the U-boat was out of wireless range – as they hunted for easy prey around the British Isles.

However, because of Room 40's isolation from other sections of the Admiralty, and even from other departments within Naval Intelligence, it had no way of applying the knowledge it gathered, deprived as it was of any information regarding the location or trajectory of Allied shipping. Room 40 simply passed its findings on for others to analyse. Nevertheless, messages intercepted by the codebreakers from German submarines and wireless transmitting stations were passed on to the British navy, and to ships at risk, via the slightly creaky chain of command.

This particular U-boat's commander, Walter Schwieger, and his 32-man crew had gained notoriety for firing at – and missing – the *Asturias*, a British hospital ship, in February of that same year. On 5 May, U-20 sank the three-masted wooden British schooner the *Earl of Lathom*, in a grenade attack. The next day, it sank the British steamers *Candidate* and *Centurion*.

Schwieger was planning on heading back to base, but had to linger to avoid HMS *Juno*, a passing British cruiser. He had just two torpedoes left, and the *Juno* was moving fast and zigzagging. This, combined with the fog, made the prospect of a shot at her the kind of chance Schwieger didn't want to take.

And then along came the *Lusitania*, which Schwieger knew was an English ship. Her course and speed gave him a target that no amount of planning could guarantee: a clear bow shot from 700 metres. At 2.10 p.m., the single torpedo fired by Schwieger's U-20 struck the *Lusitania* just under her bridge. Schwieger later described the scene in his diary: 'A second explosion must have followed that of the torpedo (boiler or coal or powder?) … The ship stopped immediately and quickly listed sharply to starboard, sinking deeper by the head at the same time. It appeared as if it would capsize in a short time. Great confusion arose on the ship; some of the boats were swung clear and lowered into the water.'

The second explosion was the source of much outrage, another example of German overkill. However, the fatal blow was struck by U-20's torpedo; the secondary explosion was likely the result of a chemical chain reaction in the *Lusitania*'s cargo. The result, though, was lethal to more than half of the 1,959 passengers on board. In the end, 1,196 died – 128 of them American citizens, the majority of whom were women and children.

The death of American women and children was trumpeted by the Allies, in case the Americans had been unmoved by this latest, and to date worst, German atrocity involving the slaughter of innocents. In London, the newspapers called the *Lusitania*'s sinking a massacre in cold blood. Across the United States, the press followed the story first with hope that all had survived; the 7 May edition of the *Tacoma Times* of Washington State, enjoying an eight-hour time difference from the site of the sinking, claimed that 'latest reports say that all persons on board were saved by lifeboats'. The next day, the reports were angry and sad, as details came into focus: 'Germany Glad Ship Sunk: 1,200 Die' shouted a banner headline on the *El Paso Herald*, adding beneath it, 'Weeping Widows Mourn Dead'.

Would this be the fatal blow to bring the United States into the European war on the side of the Allies? Former President Theodore Roosevelt, who made no secret of his desire to get America into the fight, said, 'It seems inconceivable that we should refrain from taking action on this matter, for we owe it not only to humanity but to our own national self-respect.' From the White House, however, there was a disquieting silence.

There was nothing but noise from the German-American community. At Lüchow's popular German restaurant near Union Square in Manhattan, exuberant German families enjoyed the orchestra's patriotic German songs, while a few blocks north, at Hofbrau Haus, patrons were raising toasts to the Kaiser and to the sinking of the *Lusitania*. At the exclusive German Club, one cavalry captain, stranded in New York by the war, said, 'This is a masterstroke which will curb transatlantic travelling and isolate Great Britain more effectively than a whole fleet of super-dreadnoughts could possibly accomplish. It's the doom of Great Britain.'

And yet President Woodrow Wilson was still silent. Wilson had just finished his Friday lunch and was preparing for a round of golf when he received news of the *Lusitania*'s sinking via a telegram from the US Consul in Cork, Ireland on 7 May. The first report said all souls on board were safe, but still, Wilson cancelled his golf game and went for a drive. He had other things on his mind, the most pressing being his wooing of the widow Edith Galt, whom he had met in March. 'My happiness absolutely depends on your giving me your entire love,' he wrote to her that night, while news reports now painted a more complete picture of the carnage on the Irish Sea.

The outrage was surely enough to provoke the US to declare war on Germany, but Wilson's reluctance to go to war was born partly by his desire to win the peace in Europe, but more by the political and social volatility of his own country. Twelve million immigrants had landed in the USA since the turn of the 20th century, profoundly affecting a country of 100 million people. Labour reformers clashed with business interests after the financial crisis of 1907 as both attempted to wrest control of the social marketplace, and the migration of millions of African Americans from the southern US to the north created social upheaval compounded by the immigrant tidal wave as people competed for jobs and housing. And then there was Mexico, embroiled in a revolution, on the USA's southern border. Wilson feared the effects of declaring war on Germany on the stability of the country of which he had been president for less than two years.

Wilson's first public response to the *Lusitania*'s destruction and American loss of life came on 10 May in Philadelphia, where he addressed 15,000 people at the Convention Hall – 4,000 of them newly naturalised American citizens. 'There is such a thing as a man being too proud to fight,' he told the crowd. 'There is such a thing as a nation being so right that it does not need to convince others by force that it is right.' The crowd, waving thousands of small American flags, erupted in a tumult of applause and patriotic enthusiasm, for even though Wilson had not mentioned the *Lusitania* (nor would he in that speech), 'the audience did not hesitate to read the application of his statement'.

Count von Bernstorff, the German ambassador, had endured both a bomb threat at the German embassy in Washington, and the persistence of a group of US newspaper reporters, who chased him in a speeding taxi from his suite at the Ritz-Carlton to Penn Station. There the cornered von Bernstorff fended off their questions in surreal fashion, first by claiming that he wasn't there at all, and then, in relief, agreeing to a reporter's suggestion that he had to wait for instructions from his government.

But what really mattered to the world was America's response. In Berlin, American ambassador James Gerard fully expected his country to declare war on Germany, and prepared to depart for home. Instead, on 11 May, he found himself delivering the first American '*Lusitania* Note' to the Germans. The Wilson government's high-minded response did not declare war, but rather asserted that Germany should not expect the United States 'to omit any word or any act necessary to the sacred duty of maintaining the rights of the United States and its citizens'.

Gerard had frequent conversations during that intense post-*Lusitania* period with Germany's Foreign Minister Gottlieb von Jagow and Under Secretary of State Arthur Zimmermann. On one heated occasion Zimmermann, pounding the table with his fist, told Gerard that the United States wouldn't dare to retaliate because 'we have five hundred thousand German reservists in America who will rise in arms against your government'. Gerard's response was highly undiplomatic, but traditionally American in its use of mob violence: 'I told him we had five hundred and one thousand lamp posts in America, and that is where the German reservists would find themselves if they tried any uprising.'

Germany apologised to the United States for the loss of the *Lusitania*, but never accepted blame, arguing that it had destroyed the ship within the rules of warfare because it was carrying war goods. Even so, the German government wanted to keep the United States from joining the Allied cause and issued secret orders to its submarine captains to stop sinking passenger liners.

The United States accepted the German government's apology. America would not be going to war with Germany over the *Lusitania*, much to the dismay of the Allies, who now wondered just what sort of outrage it would take to bring her into the fight. And the Germans didn't need their 500,000 reservists to rise up in America. Von Bernstorff and his masters in Berlin were exploring other, potentially more deadly, options.

# Chapter 4

# MILITARY MATTERS

When a German machine-gun bullet struck Malcolm Hay in the head, he knew instantly what had happened: 'the blow might have come from a sledge hammer, except that it seemed to carry with it an impression of speed'. In the few seconds before he lost consciousness, he noticed that his watch was spattered with blood. When he came round, he was 'unable to move any part of my body except my left hand'.

Hay, future head of the army's codebreaking outfit MI1(b), had joined the Gordon Highlanders on the day war was declared, and arrived in France as part of the British Expeditionary Force (BEF) in August 1914. As he marched through the countryside towards the Belgian town of Mons, enjoying a late summer heatwave, there wasn't 'the slightest hint of war'. That would rapidly change: his unit was about to bear the full brunt of the massive German offensive designed to drive the BEF back where they came from and deliver a devastating right hook towards Paris. Hay noted that 'the German superiority at that part of the line was probably about three to one in guns, and five or more to one in men'.

Besieged from the flanks and the rear, Hay and his troops joined the general British retreat; suddenly the glorious weather was their enemy as they trudged along the endless road, suffering from the choking dust and the hot sun. Near the small village of Bethancourt, Hay called a halt and they dug in. Heavy shelling began the next morning, followed by wave upon wave of German infantry. As they swarmed forward, Hay was struck down.

For the rest of the day, he swam in and out of consciousness, tended to by his men. Ordered to retire at midnight, they carried him away on a stretcher, the pain almost unbearable: 'I still remember the agony caused by the weight of my body pressing down on my neck ... while my head, just clearing the ground, trailed among the wet beetroot leaves.'

The next day, further orders arrived: the wounded were to be left behind. Two of Hay's men stayed with him as long as possible. Then he lay helplessly by the roadside until the Red Cross picked him up and he was ferried to a civilian hospital at Cambrai. For the first month, he was attended to by French medical personnel, who were struggling to accommodate the flood of Allied casualties filling every available space: wounded men were lying in the corridor.

His condition remained critical: an abscess had formed in the wound owing to the presence of a bone splinter, and he needed surgery. The doctor used a lancet to reveal the jagged, splintered edge of the skull. Then he 'broke off one or two pieces of bone about the size of a tooth, then jammed in a piece of lint soaked in iodine'. Hay survived this crude procedure but was still too sick to be moved, whereas most of the captured British casualties, victims of the same massive German offensive that had netted Hay, were being 'sent to Germany packed in cattle trucks, with no medical attendance, no food, no water'.

On 21 October, he was transferred to a small, gloomy French Red Cross facility in an old school building. Here his knowledge of the language allowed him to develop friendships with the locals who visited the sick, including a woman who would cook them tasty meals that Hay would help prepare, peeling the potatoes for the twice-weekly treat of *pommes frites*. He even conquered his aversion to snails. Another bonus was a consignment of English tobacco courtesy of a patriotic *marchande de tabac*, who had buried the most valuable part of her stock in a back garden.

As he slowly recovered feeling in his stricken limbs and began to gingerly walk again, Hay observed the consequences of German occupation: such things as bicycles and sewing machines were requisitioned, while those caught in possession of pigeons – which were used by both sides to carry messages – would be condemned to death.

Once the Germans took charge of the hospital, conditions deteriorated rapidly, the wounded piled up, and Hay was kept awake at night by the mournful cries of the dying. He also received a stark reminder of how lucky he'd been: 'a French soldier was brought in with a head wound in almost exactly the same place as my own, but a hair's breadth more to the front of the head. This difference of perhaps a tenth of a millimetre had left the Frenchman deprived of speech, memory and motion.'

Shortly after Christmas, which was marked by a tree, turkey and plum pudding, Hay joined the list of prisoners deemed transportable and left by train on 6 January. The journey seemed interminable. There were several changes, food was a rarity, and once in Germany itself, Hay and his fellow prisoners were subjected to abuse from their guards – who 'behaved with great rudeness and barbarity' – and from hostile crowds gathered on station platforms.

Eventually he arrived at his destination, Festung Marienberg, 'a place of great architectural and historical interest', about a mile outside the town of Würzburg, and joined around 50 fellow officers, mostly French, with a sprinkling of British prisoners. Hay slept on a narrow wooden plank with a mattress made from 'a coarse linen sack ... stuffed with straw'. His few privileges included being able to write one letter a day; a weekly visit to Würzburg, where he was allowed to buy food to supplement his meagre cabbage-based diet; and a wash in the public baths once a month.

In his spare time he played cards, chatted with his fellow inmates, attended Mass and worked on building up his strength. Aside from the bitter cold of winter, boredom was the main discomfort: 'the misery of inactivity ... the monotony, the aimless futility of existence ... this is the real trial that makes prison intolerable'.

From day one of his captivity, he was looking for a way out. If he could convince the authorities that his injuries were serious enough to rule out any possibility of him fighting again, they might agree to release him. Though his medical certificate, 'a most alarming piece of evidence as to my condition', supported his case, he also appealed to the American ambassador in Berlin, and to a very well-connected family friend, Princess Blücher. She wrote to him on 29 January promising to 'do my best to get you included among those for exchanges'. She handed his medical report to the American consul, who was staying at her hotel, and he agreed to do what he could to help.

Her timely intervention sealed his fate, and on 12 February 1915, his doctor told him he was going home. By the 16th, having travelled first class and been treated 'with all possible kindness', he was in Holland waiting for a boat back to England, his mood strangely muted; 'not hilarious excitement or placid contentment but an exceeding weariness of mind and body'.

The whole experience left him convinced that Germany was in the grip of a virulent militaristic nationalism that threatened everything he held dear. Had Malcolm Hay's captors known what a significant contribution he was going to make to the Allied war effort, they would never have let him go.

Tall and slim, with blue eyes, Malcolm Hay was born on 21 January 1881 into an aristocratic Scottish family with connections to medieval French nobility and a 1,000-acre estate at Seaton, near Aberdeen. His parents played only a minor role in his childhood: his father was an old man, while his mother tragically died of diabetes when Malcolm was 11.

At school, he showed a gift for languages and considerable sporting talent, playing golf, tennis and hockey. Aged 17, he was packed off to stay with relatives in France. Life at their country chateau was pleasant enough, but Malcolm professed not to have enjoyed his time there and resented missing out on higher education. In 1902, he married his first cousin, Florence de Thienne, and the couple had two children. In 1908, he returned to Scotland and assumed control of his ancestral lands. A conscientious and benign estate manager, he got to know every one of his tenants personally.

Though Hay sailed, fished and hunted, he did not conform to the stereotype of a country Lord: he was politically liberal and a Scottish nationalist. In 1900, just two years after it was founded, he joined the

Royal Automobile Club (RAC), an extremely exclusive organisation given the sheer expense and novelty of owning a car. Membership of the RAC brought with it a host of perks: repairs done on the cheap by specially appointed mechanics, reduced rates at certain hotels, and a dispensation from the French government that allowed members to take their vehicles into the country without going through customs.

On the surface, Hay and Blinker Hall could not have been more different. Alice Ivy Hay, Malcolm's second wife, wrote that Malcolm greatly admired Hall but thought 'he could be entirely ruthless, especially in what he considered to be the execution of his duty'. By contrast, Hay was a thoughtful, compassionate man 'with a genius for friendship'. Yet his gentleness concealed a hard inner core: 'he could be fierce too, especially in his hatred of injustice'.

It was his mental and physical strength that got him through the ordeal of incarceration and meant he shrugged off the wounds that left him permanently disabled. To keep fit, he rode a tricycle round London, much to the dismay of his family, and he refused to let his lame left leg prevent him from playing golf: 'he invented and used a weird and wonderful golf putter, a kind of cross between a croquet mallet and a croupier's rake'.

On his return from Germany, and after a period of convalescence and rehabilitation at his manorial home in Scotland, Hay headed for London, where he presented himself at the War Office (WO) and 'begged to be given something to do – otherwise, he insisted, he would die'. He was fluent in French and Italian, and had a good knowledge of both German and Spanish, and since the War Office habitually used wounded officers for intelligence work, he was offered the job of running MI1(b), the military's cryptographic outfit. He accepted on the condition that there would be no interference from the WO and he would have a free hand to run MI1(b) as he saw fit. Naturally the mandarins resisted his demands for as long as possible but relented when he threatened to quit.

Though he had no previous experience, Malcolm Hay proved to be a natural codebreaker. As long as there was a sufficient quantity of messages to act as a guide, he believed any code could be solved.

Rigorous and pragmatic, he thought that 'anyone of average intelligence can learn to read cryptograms', on the basis that 'every message, every sentence in a message, has its own distinctive pattern, and contains repetitions of letters, of combinations of letters, and sometimes of complete words', and that these repetitions were especially vulnerable to frequency analysis: observing how often and in what position letters appeared in the text. However much 'these patterns can be disguised, they cannot be destroyed'.

Hay identified two main methods of disguise: substitution and transposition. In the first, 'the letters of the text are replaced by various symbols', usually letters or numbers. In the second, 'the letters of the text change their position according to some predetermined arrangement, or key'.

The 'key', generally known as the cipher key, was composed of numbers or letters, sometimes randomly organised, sometimes not – ciphers often featured names, or common phrases, or snippets from songs or poems – of varying lengths, that were attached to the beginning of the encoded message and determined what arrangement of the alphabet was being used in the rest of the text. Uncovering the formula of the cipher key, a device used by the German navy, its diplomats and secret services, was an essential part of the codebreaker's work.

The man Malcolm Hay replaced as the boss of the military's codebreakers at MI1(b) was Brigadier General Francis Anderson, who had successfully analysed the enemy's ciphers during the Boer War and had written several pamphlets on the subject of codebreaking. Though retired at the outbreak of hostilities in 1914, Anderson didn't hesitate to offer his services as the Allies fought to withstand the massive German onslaught that the Kaiser hoped would end the war quickly.

Up to that point, although the War Office had recognised the need to develop its cryptographic capabilities, it hadn't done much beyond publishing a manual and making tentative efforts to examine German army field ciphers. As the BEF arrived in France, it set up wireless interception facilities that soon picked up a vast amount of traffic, both German and French. However, reception was poor and the messages were often garbled and full of errors.

When it came to the vital first few months of the conflict, it would be the French who led the way. Their cryptographic unit, the *Bureau de Chiffre*, had been established in the wake of the Franco-Prussian war of 1870, when the French were humbled by Bismarck's army, to act as the first line of defence against further aggression. As the Germans massed once again on their borders, the French had ten wireless interception stations ready to receive the enemy's messages, including one at the top of the Eiffel Tower. The material coming in from these sources enabled the Bureau to break the German codes. So efficient were the French codebreakers that when the Germans introduced a new cipher, it defied the Bureau for only three weeks.

By that time, the Germans' mammoth offensive had run out of steam. Their exhausted wireless operators, overwhelmed by the sheer volume of material they had to transmit, had resorted to broadcasting *en clair*. The Allies picked up these uncoded messages, giving them the information they needed to repulse the enemy assault at the First Battle of the Marne in early September, and plan the subsequent counter-attack that would drive a wedge between the German armies, forcing them to retreat.

Over the next couple of months, more *en clair* communications warned the BEF about six German attacks as both sides raced towards the French coast. Had the Germans got there first, the BEF would have been stranded, cut off from supplies and reinforcements. Instead, it was able to dig in and establish a defensive line. The Germans did the same. Four years of trench warfare had begun.

At MI1(b), Hay inherited three members of staff, including an unemployed 40-year-old aesthete, Oliver Strachey, who would become one of the key figures in Britain's codebreaking apparatus. A man of abundant gifts, Strachey lacked the motivation to capitalise on them until he became a codebreaker in 1914 and found his true calling.

A confirmed atheist, he enjoyed life and viewed it as a game not to be taken too seriously. A friend of his daughter described him as 'gregarious, amused, amusing, highly intelligent, and interested in everything'. His subversive nature is best captured by an anecdote concerning his time as a juror on what appeared to be an open-and-shut case. A well-known robber was being tried for loitering with intent outside a house with a crowbar down the inside of his trouser leg. With the rest of the jury ready to rubber-stamp a guilty verdict, Strachey, out of 'sheer devilry', thought it would be amusing to see if he could change their minds 'by force of logical argument'. Though it took some doing, he persuaded them that the thief was not guilty. Back in court, when the foreman delivered their shock decision, 'the judge's mouth fell open – but he wasn't nearly as surprised as the prisoner'.

Strachey's non-conformist character owed a lot to his mother, Lady Jane Strachey. A modern woman – she smoked and played billiards – with a passion for literature, liberal politics and feminist causes, she passed on her love of words to her children and encouraged them to let their imaginations run riot. Oliver's younger brother Lytton found literary fame as the author of *Eminent Victorians* (1918); his sister Dorothy translated the writings of

*Oliver Strachey, leading codebreaker, lost in thought*

her friend André Gide into English; while another brother, James, would do the same for the works of Sigmund Freud.

Educated at Summerfield and Eton, where he excelled at languages, mathematics and music – the piano was his first and most enduring love – Strachey managed to get himself expelled from Balliol College, Oxford, after just two terms. Rumours that he'd had a homosexual affair were dismissed as absurd by all who knew him: he was enthusiastically heterosexual. Having failed to make the grade as a concert pianist in Vienna, he was packed off to the Raj, where he worked as a traffic superintendent on the East India Railway. Thoroughly bored, he met and married a Swiss-German beauty, a relationship that soon ended in acrimonious divorce.

Back in England, Strachey mixed effortlessly with the Bloomsbury Set, that decadent anti-establishment group who left a major imprint on twentieth-century culture. The writers Virginia and Leonard Woolf and the artists Duncan Grant, Roger Fry and Vanessa Bell would become lifelong friends. Fancy-dress parties were a regular fixture: at one, Strachey was the Harlequin; at another he came as the ballet dancer Nijinsky, dressed all in red. During long weekends in the country, he would spend hours debating ethics with the moral philosopher G. E. Moore, performing music, and playing marathon games of chess and bridge, two of his favourite pastimes.

Around this time, he met the genius codebreaker Dilly Knox, who was then deciphering ancient Greek poetry. Lytton Strachey, Oliver's brother, was a contemporary of Dilly's at Cambridge and developed a huge crush on him, which was not reciprocated. Oliver would bump into Dilly at Bloomsbury gatherings, and over the course of their long codebreaking careers they became firm friends.

In 1911, Strachey married Ray Costelloe. Known for her unflattering dress sense – her grandmother observed that 'Ray sniffs at the idea of trying to make her look graceful … a hopeless task … but she has consented to give up those awful knickers for the summer' – she was distinctly unfeminine and loved sport and fast cars. She read mathematics at Newnham College, Cambridge, and after graduating met Pippa Strachey, who was deeply committed to the women's movement and worked for the Central Society for Women's Suffrage. Ray joined and threw herself into organisational duties and campaigning. She would play a pivotal role in mobilising Britain's women during the war.

After honeymooning in India, the newly-weds returned to London in early 1912. They moved into a tiny flat and lived off an allowance from Ray's stepfather. The opportunity for Strachey to join MI1(b) came through a family friend at the War Office, who was 'looking for someone with an ingenious head for puzzles and acrostics to decipher code and piece together scraps of wireless messages'. A meeting was arranged and he was hired on the spot. With his agile brain, his knowledge of languages, mathematics and music, his skill at games and puzzles, coupled with his love of detective novels that tested his ability to find clues and solve the mystery, he fitted in perfectly.

When Malcolm Hay arrived at MI1(b) in late summer 1915, work on Western Front material had slowed to a trickle: the Germans had for the time being stopped using wireless and were sending their messages via trench telephone. Instead, Strachey and his two compatriots were attempting 'the reconstruction of the American diplomatic code books'. Without economic and financial support from America the Allies could not continue fighting. Knowing the intimate details of US policy could make all the difference between winning and losing. Hay immediately set about examining communications between James Gerard, the US ambassador in Berlin, and his government in Washington. According to Alice Ivy Hay, his second wife, Malcolm detected their meaning over the course of one night.

Without a steady stream of wireless intercepts to work with, Hay next turned his attention to the voluminous telegraph traffic passing through London. He discovered that copies of all foreign telegrams were kept by the cable censor, Lord Arthur Browne, who was a member of the War Office staff. It was thus possible to obtain copies of all the diplomatic cables which passed through London.

Up to this point, the idea that the confidential communications of neutral and friendly powers constituted fair game was anathema to the

MILITARY MATTERS

British establishment. Such behaviour was considered ungentlemanly, not worthy of an Englishman. Collecting them was one thing, reading them quite a different matter. Hay, however, was not bothered by etiquette. There was a war to win. He quickly realised the potential value of the enormous mass of encoded messages from all over the world which were accumulating in War Office cupboards. He met with Browne, who agreed to have all diplomatic cable traffic handed over to MI1(b).

Hay was also determined to rejuvenate MI1(b)'s relationship with Room 40. During the early months of the war, as both organisations were finding their feet, they worked in relative harmony. Alastair Denniston, one of Room 40's first recruits, remembered how work on the ciphers continued in the Admiralty and the War Office by day, while the night watch worked in the War Office. However, this cordial *esprit de corps* soon broke down, as the long-held rivalry between the Admiralty and the War Office reasserted itself. As a result, 'a definite breach' occurred, and from then on the two sets of codebreakers worked in isolation.

Hay immediately recognised the need for them to cooperate and informally approached Blinker Hall. Never one to let institutional loyalty interfere with beating the Germans, Hall was happy to oblige, especially as Room 40 had also begun to deal with the Germans' diplomatic communications, thanks to a bizarre chain of events in the Middle East.

Chapter 5

# SPIES IN AMERICA

When the Ottoman Empire entered the war in 1914, it brought with it hundreds of thousands of Muslims, the holiest sites in Islam and the supreme leader of the faith, the Sultan-Caliph Mehmed Rashad V. The Germans sensed an opportunity to destabilise the British Empire by appealing to its millions of Muslim subjects in India, North Africa and the Far East and rallying them to their standard.

A major step in that direction was taken when the Sultan-Caliph declared a jihad against the British in November 1914. An intensive propaganda campaign was launched by German agents, underground networks were established in Egypt, and a small, dedicated team was dispatched to foment uprisings in Persia (Iran), notionally independent but jointly controlled by Britain and Russia, and Afghanistan.

One of these insurrectionists was Wilhelm Wassmuss. The nearest German equivalent to Lawrence of Arabia, Wassmuss was a career diplomat, fluent in Persian and Arabic, who adopted the lifestyle of the desert tribes. An advocate of guerrilla warfare, he was single-handedly responsible for leading the British a merry dance in southern Persia: stirring up locals, capturing several towns, blowing up oil pipelines and taking hostages. An irritant rather than a major threat, Wassmuss inadvertently gifted Room 40 the code books that would ultimately transform Allied fortunes.

By the spring of 1915, the British were hot on his heels, but Wassmuss always managed to keep one step ahead, twice escaping the somewhat lacklustre efforts of local pro-British officials to hold him prisoner. At this point, history and myth collide. According to one account of events, when Wassmuss eluded his captors he left behind a chest containing secret papers, amongst them two versions of the German diplomatic code book, numbered 89734 and 3512. The British seized the trunk and it made its way to the India Office in London (Persia was under the jurisdiction of the Raj). Alternatively, the British, annoyed that Wassmuss had got away, arrested the German consul in the Persian Gulf for collaborating with him. During a search of the consul's office safe they came across the code books. Not appreciating their potential value, they sent them back to London.

By chance, Blinker Hall bumped into a young naval officer who told him about the Wassmuss affair. He wasted no time getting the books out of storage. Alistair Denniston remembered how one day in April Hall 'produced a fresh line of goods – a treasure trove in Persia'.

Hall immediately formed a diplomatic section that was independent from the rest of Room 40, giving him sole control of its activities. He head-hunted George Young, a very experienced Foreign Office official who'd worked in Washington, Athens and Constantinople, and was based in Lisbon at the outbreak of war, and placed him in charge. To make up the rest of the team, he pinched Nigel de Grey and the Reverend Montgomery from Room 40, and brought in Benjamin Faudel-Phillips, a City man.

The diplomatic code books yielded crucial intelligence concerning communications between Berlin and the German embassy in Madrid, which was acting as a clearing house for messages to their spies in America, where an extensive sabotage operation was being planned: before the code book was discovered, 170 German messages had passed via neutral cables to Count Johann von Bernstorff, the larger-than-life German ambassador in America.

However, because of the risks of having their ambassador oversee a covert war on US turf, the Germans provided von Bernstorff with plausible deniability in the form of like-minded associates. The German embassy had an executive staff of four, who would in effect become the general command of the German war effort in North America, and have at their disposal an army of lethal patriots and rogues who saw the US financing and supply of the Allied war effort as a legitimate reason to wage war on America.

LIBRARY OF CONGRESS, WASHINGTON

*Heinrich Albert, the German commercial attaché*
*in the USA and sabotage 'money man'*

The first was Germany's commercial attaché in the US, Privy Councillor Dr Heinrich Albert, a 40-year-old lawyer. Albert was paymaster for the German diplomatic corps in the US – and the eventual paymaster to Germany's espionage and sabotage, holding a massive joint account with Ambassador von Bernstorff at Chase National Bank. He was popular with his American banking colleagues, and conveyed an air of discreet competence despite the vicious duelling scars that creased his face. From his office deep in New York's financial community, at the Hamburg America Building in Lower Manhattan, Heinrich exerted great influence with New York bankers as a man of prudence and principle. Yet in truth his wartime activities for the Fatherland characterised him, in a later Senate investigation, as 'the Machiavelli of the whole thing … the mildest mannered man that ever scuttled a ship or cut a throat'.

The German military attaché in the US was also an aristocrat, Captain Franz von Papen. The eldest son of a wealthy, noble and Catholic land-owning family in Westphalia, the 35-year-old von Papen was an officer of an Uhlan cavalry regiment, and had recently finished a stint as a military attendant to the Imperial Palace when he was dispatched as attaché to the United States and Mexico in 1913. Tall, powerfully built and with a sculpted, strong-jawed face that gave him an air of both vigour and sneering arrogance, he, like von Bernstorff, had married money. His wife's fortune, as the daughter of an Alsatian pottery manufacturer, gave von Papen social standing in Washington DC, which he used to pursue other women for reasons personal and political, though he was more often than not to be found in his redoubt at 60 Wall Street in the heart of New York's financial district, which became known as the Bureau of the Military Attaché – or more nakedly, the War Intelligence Centre.

*Franz von Papen, Germany's military attaché in the USA*

Rounding out the quartet was Germany's naval attaché, Captain Karl Boy-Ed, who had joined the navy at the age of 19 and seen action around the world. He witnessed the brutal American invasion of the Philippines in 1898, and was a secret agent for the Kaiser's brother shortly before the Boxer War in China, where his mission was to measure the strength of the Chinese navy. Boy-Ed's origins were more exotically bohemian than those of his von-prefixed colleagues in Germany's American war office. He was born in the important (and intellectually vibrant) German seaport of Lubbock, on the Baltic coast, where his father was a merchant of Turkish ancestry, and his progressive, intellectual mother was a journalist and novelist who nurtured the career of Thomas Mann, a frequent guest in the Boy-Ed household.

Tall and built like a rugby prop, Boy-Ed was worldly, well read and funny, with a compelling combination of charm and diligence that made him popular among the Washington crowd when he took up his diplomatic post in 1913. When war came in 1914, he too set up his office in New York City, at 11 Broadway, close to the New York Customs House. And like von Papen, he was within easy walking distance of Heinrich Albert's counting house.

New York City was the perfect North American front line for the Germans' secret war. With a population of 5.3 million people, it was the largest city in the world. Better still, many of the 12 million immigrants who had landed on Ellis Island between 1900 and 1915 stayed where they'd arrived. About one million residents of New York City were foreign born, largely Irish and German, with no love lost for England.

And the city itself was a marvel of modernity. At 7.30 p.m. on 24 June 1913, President Woodrow Wilson pressed an electric switch in the White House, and on the corner of Broadway and Park Row in Manhattan, 200 miles up the road, the 80,000 lights of the new 60-storey Woolworth Building blazed out to ships 40 miles out to sea that the world's largest skyscraper was open for business in a city whose new epicenter – Times Square – called itself the crossroads of the world. Ezra Pound, the American poet then living in London, visited the city and was moved to remark, 'No urban nights are like nights there. I have looked down across the city from high windows. It is then that the great buildings lose reality and take on magical powers. Squares and squares of flame set and cut into the ether. Here is our poetry, for we have pulled the stars down to our will.' With 500 miles of shoreline, the city's five boroughs were connected to each other and the rest of the country by ferry, tunnel and bridges spanning a harbour that was the busiest in the world, funnelling goods into the American heartland and out to the world beyond.

Von Papen's first instinct, however, was to look further afield and attack Canada, which, as a self-governing dominion of Great Britain, had gone to war along with Britain and was now a vital source of soldiers and supplies for the war in Europe, sending 30,000 soldiers to England in October 1914 – 5,000 more than London had rather extravagantly hoped for in such a short time. Canada's military training centre at Valcartier, Quebec, 500 miles north of New York City, was within easy striking distance, but for the German saboteurs in the USA, the entire country, as a key part of the British Empire's war machine, was a target.

The Canadian attack plan had come to von Papen via a German soldier of fortune named Holst von der Goltz, who had attained the rank of major while serving with Pancho Villa's revolutionary army in Mexico. The baby-faced von der Goltz was really Franz Wachendorf, born in Koblenz in 1884, who already boasted a shadowy career in intelligence, having stolen a document from a Mexican finance minister he had chloroformed in Paris. The purloined document revealed a top-secret agreement between Japan and Mexico, which von der Goltz then leaked to the US in February 1911, who responded by sending their fleet to the Gulf of Mexico, and 20,000 troops – two-thirds of the US army at the time – to the Mexican border.

The following year von der Goltz was in Mexico too, fighting the revolution. That adventure ended once Germany was at war. He heeded the call for all German soldiers and reservists located outside Europe to muster in the United States, and headed to New York to pay a call on Franz von Papen.

On 22 August 1914, von der Goltz appeared at 11 Broadway, and in a little room in the Imperial German consulate he and von Papen hatched a plan to strike at Britain through its dominion of Canada. Von Papen produced a letter he had received from a German who worked on a farm in Oregon suggesting that the Germans should attack Canada's Great Lakes cities with machine guns mounted on motorboats. The two men agreed that the plan was good, but that it could also be a trap, and so von Papen, with the same kind of shrewd thoroughness that would later see him rise to vice chancellor of Nazi Germany under Adolf Hitler, had the letter-writer investigated.

In the meantime, the duo hatched a scheme to use German reservists in the United States to invade British Columbia via Washington State, with support from German warships in the Pacific. That plan was rejected too, partly because of the lack of artillery backup, but mainly because Ambassador von Bernstorff thought it the kind of 'wild-cat scheme' that would inflame Germanophobia and ultimately fail.

After next making serious plans to attack the British Empire by invading the soft target of Jamaica, which would soon send men and supplies to Europe once the appalling Allied casualties on the Western Front overcame British military prejudices of giving black men guns, von der Goltz and von Papen decided instead to blow up the Welland Canal, a major Canadian shipping route between Lakes Ontario and Erie that bypasses Niagara Falls. A successful attack on the Welland Canal system would cripple Canadian shipping and food supplies and create panic among the public, who would demand that troops destined for the war in Europe be stationed at home to prevent invasion by the German horde to the south.

Von der Goltz was promoted to captain and acquired a US passport in the name of Bridgeman Taylor as well as 300 pounds of dynamite and 45 feet of fuse. He purchased the explosives package through the offices of Captain Hans Tauscher, an agent of Krupp, the German steel and armament juggernaut, via the DuPont Powder Company, for $534.37.

After picking up the dynamite and fuse himself by motorboat from a DuPont company barge in the Hudson River, von der Goltz transported it in suitcases in a careening New York taxi first to the German Club on the south side of Central Park, where Karl Boy-Ed lived, and which became such a centre for German diplomats, sympathisers and saboteurs that the

US government finally seized the building in 1918. From there he went to the safe house run by Martha Held, a buxom, blue-eyed opera singer who had emigrated from Germany in 1912. With her raven hair, diamond earrings and the elaborate Victorian costumes that camouflaged her middle-age spread, Held trilled arias into the Manhattan night from her rented brownstone at 123 West 15th Street, an easy address to remember for the German sailors, officers and spies – and even von Bernstorff – who were regular visitors.

Once the explosive was safely stowed, von der Goltz and his hand-picked team of German sailors whom he had liberated from their involuntary residence in American ports headed north to strike a blow for Germany. However, not only had the Canadians figured out that such an important shipping lane might be a target for saboteurs and guarded it accordingly, by the time von der Goltz and his fellow conspirators arrived in Niagara Falls, New York, they learned that Canadian troops had left their base at Valcartier for England. So instead, they exchanged impotent telegrams with Mr Steffens (von Papen's code name) and talked loudly about their German patriotism for all and sundry to hear. Von der Goltz claimed that it was to distract their American surveillance team from the other unit of German saboteurs heading into Canada, but the truth was that the Germans were going to have to come up with a better plan than this if they hoped to win the war on the North American front.

On 3 January 1915, an enciphered telegram arrived at Ambassador von Bernstorff's Washington office, with orders from Berlin:

> Secret. The General Staff is anxious that vigorous measures should be taken to destroy the Canadian Pacific in several places for the purpose of causing a lengthy interruption of traffic. Captain Boehm who is well known in America and who will shortly return to that country is furnished with expert information on that subject. Acquaint the military attaché with the above and furnish the sums required for the enterprise.

The sender of the telegram was Arthur Zimmermann, then the Under Secretary of State in Germany's Foreign Office, who would send a much more incendiary telegram to America two years later. For now, Berlin was worried that the Canadian Pacific Railway was going to transport Japanese troops across Canada and on to ships to deliver them to the Allied cause in Europe, which Japan had joined at the end of August.

Von Bernstorff passed the order on to von Papen, who went hunting for a suitable candidate to blow up the Canadian Pacific. He found him in Werner Horn, a former lieutenant in the German army, who had been the manager of a coffee plant in Guatemala when he heard the bugles call him back to war. Making his way north, Horn reached New York City and reported for duty.

The 37-year-old reservist was not by nature a saboteur, nor a cold-blooded soldier, though he was extremely cold as he tramped toward his target through the frigid midnight air in the early hours of 2 February 1915. Horn had been paid $700 for his sabotage services by von Papen, and kitted out with dynamite, which he carried in suitcases. Von Papen had told him that by destroying the deserted McAdam-Vanceboro railway bridge linking Maine with New Brunswick in the middle of the night, he would prevent lethal munitions from killing his countrymen on the battlefield. And if he pinned the German colours to his coat, he would be treated as a soldier and not a spy if caught.

The guileless Horn wanted to serve the Fatherland, but he wasn't keen on killing anyone, nor on damaging property in the United States. So with a harsh wind chill dropping the mercury to minus 30°F, he hauled his dynamite to the Canadian side of the bridge. He was nearly run over twice by passing trains despite the timetable von Papen had given him that said the bridge would be unused in the wee hours of the night. Given his desire to avoid death – including his own – he cut the 50-minute fuse to three minutes, reckoning that no train would pass by in the brief interval, and that the explosion would alert the townspeople to stop the next one. Then he lit the fuse with his cigar and, with frostbite attacking his hands, feet, nose and ears, hurried back to the hotel from which he'd recently checked out.

The explosion knocked out windows and sent the freezing Arctic air into the bedrooms of Vanceboro. But all it did to the bridge was twist the steel girders, which merely rendered it unusable for a few days. Horn was soon apprehended when the manager of the hotel reported that his 'Danish' guest had returned with frostbite. He was arrested and, ever the honourable soldier, admitted to everything except the identity of the man who had lied to him and sent him on the mission: Military Attaché Franz von Papen.

It had all happened with the blessing of von Bernstorff, who, despite denying any knowledge of the event until he read of it in the newspapers, sent an enciphered telegram to Zimmermann in Berlin:

Most Secret, 11 February 1915. The carrying out of your telegram No. 386 for military attaché was entrusted to a former officer, who has been arrested after [causing] an explosion on the Canadian Pacific Railway. Canada demands his extradition. I request authority to protect him; according to the laws of war, the decision ought presumably to be non-extradition, provided that an act of war is proved. I intend to argue that, although the German government has given no orders, the government regarded the causing of explosions on an enemy railway as being, since it furthered military interests, an act of war.

Zimmermann agreed, and Horn's legal expenses were covered by the German war treasury. Horn was sentenced to 18 months in a federal penitentiary in Atlanta for transporting explosives, then extradited to Canada in 1919 and sentenced to another ten years. He was judged insane in 1921 and deported to Germany.

Berlin, however, was worried. Since the war began they didn't have much to show for their North American front. They had failed to blow up the Welland Canal and the Vanceboro Bridge. They had engineered an incendiary fire on 1 January 1915 at the John A. Roebling Company plant at Trenton, New Jersey, which was famed for its wire cables for bridges such as the Brooklyn. Two days later, they created an explosion on a supply ship, the SS *Orton*, in the Erie Basin. But with the war now being fought on Western and Eastern fronts, with the Middle East soon to come, and the British naval blockade doing increasing damage domestically, it was time to up the ante.

All this nefarious activity had not gone unnoticed by Blinker Hall. The German diplomatic code books so fortuitously found in Persia had furnished Room 40 with the means to monitor the telegraphic chit-chat between the German embassy in Washington and Berlin. Decoded messages gave Hall valuable insights into the German agents' plots. Yet the use to which he could put this intelligence was limited: if he revealed the fruits of the codebreakers' labours to the Americans, he'd also have to reveal the existence of Room 40 and their interception of US communications, something he was determined not to do not least of which because of the fearsome diplomatic conflict between the US and Britain that it would undoubtedly create. He needed boots on the ground. So, as he had in other neutral countries, Hall relied on his naval attaché to act as spymaster.

In the US, this role fell to Captain Guy Gaunt, an Australian by birth, who was rated a midshipman in the Royal Naval Reserve on 17 December 1886, when he was 17 years old. He joined active service in 1895, quickly rose through the ranks, and saw battle action in the Philippines and Samoa at the end of the nineteenth century, before being promoted to captain in 1907 and becoming Britain's naval attaché in the US in 1914.

Gaunt, like other foreign service officers, found the power of New York City compared to Washington DC too great to ignore, and did most of his business from the British consul's office at 44 Whitehall Street in Lower Manhattan, just south-east of Karl Boy-Ed's HQ. Here he could monitor the traffic coming and going with the war. He liaised directly with Hall and became responsible not just for counter-espionage, but for propaganda and public relations too.

Gaunt's talents as a spy were sufficiently developed to realise that he needed help, and he found a ready supply in exiles from the Austro-Hungarian Empire, who harboured no love for the rulers of their homeland.

Chief among them was Emanuel Viktor Voska, who had escaped from Prague as a 19-year-old revolutionary in 1904, one step ahead of the police. Voska, a sculptor, went into the marble business in the USA, fathered six children, and by the time war broke out was a prosperous businessman living the American dream.

War, however, brought him back to his revolutionary roots, for he saw in the destruction of the Habsburgs the chance to establish a sovereign Czechoslovakia. Running more than 80 agents in the USA, exiles all from across the Austro-Hungarian Empire, Voska's team was able to infiltrate the German apparatus in New York, with seemingly safe clerical staff in German government offices being in the service of his network. Indeed, it was the mail clerk at the Austrian consulate who provided a list of German and Austrian military reservists who had bought fake passports to enable them to return home to join the fight.

Gaunt and Voska were not working entirely in a vacuum. They forged a strong connection with the NYPD and elements of America's fledgling intelligence services, a relationship that would result in the first major breakthrough in their struggle to foil German plans. Even though America was not officially at war, there were those within its borders who knew they were under attack. And it was time to fight back.

## Chapter 6

# THE DARK INVADER STRIKES

Between August 1914 and May 1915, the Germans sent 170 diplomatic cables to their embassy in America, discussing everything from the United States government's negotiations with Germany to German sabotage in the USA. None of these messages was processed into actionable intelligence due to Room 40's isolation from other branches of the military and the understandable caution about trespassing on America's neutral status. Blinker Hall and his team would soon engage in diplomatic surveillance, but in the meantime the Germans would ramp up their war against the Allies through their agents in the USA. From their point of view, the USA was already at war with them.

The Americans were certainly materially and financially in the battle, as ships stocked with munitions and supplies destined for the Allies steamed out of New York's 'neutral' harbour under virtual British escort – protecting them from the unrestricted submarine warfare that Germany had launched in February 1915 against Allied shipping. Franz von Papen, the German military attaché in New York, watched in impotent fury as war supplies flowed freely to the Allied cause. Ordered to report on the situation, he concluded his account to HQ in Berlin with an outraged plea: 'Something must be done to stop it.'

The German command more than agreed, especially since von Papen and Karl Boy-Ed weren't getting the job done. The man sent to intensify Germany's war in America was Franz von Rintelen, a 36-year-old captain in German naval intelligence, who set sail for America in March 1915 with

$500,000 in credit and a Swiss passport identifying him as Emile Gaché, a Swiss businessman from Bern, an identity borrowed from the husband of a fellow naval officer's sister. Von Rintelen had a trunk full of his new family's photographs – Gaché's parents' house and their alpine cottage – and the initials EVG had been sewn into his underwear, which had then been laundered to appear worn.

Franz von Rintelen arrived in New York City in fierce snow on 3 April 1915 aboard the SS *Kristianiafjord*, which had sailed from Oslo (then called Kristiania) a few days earlier. He knew the city well, for he had worked in New York a decade earlier for the influential merchant bank Kuhn, Loebe & Co. With his impeccable English and his European élan, he came across as more Italian in presentation than Teutonic, despite his pedigree as the son of a prominent German banking family. His father had been Germany's

*Franz von Rintelen, Germany's self-styled 'Dark Invader'*

minister of finance, and managing director of the Discento Bank of Berlin, and the trim, blonde von Rintelen had been expected to have a career in politics. Indeed, a seat in Germany's parliament was dangled in front of him just before he did his patriotic duty for the Fatherland and journeyed off to fight the war in America, leaving behind his political ambitions along with his wife and young daughter, whom he would not see again for six years.

New York had been good to von Rintelen during the two years he had spent in the city. The worldly and intelligent aristocrat was welcome in the highest strata of society, gracing Park Avenue dinner parties and the summer yachting season at Newport, Rhode Island; holding sway in the banking dens of Lower Manhattan and the exalted parlours of the New York Yacht Club, a Beaux-Arts bastion in midtown Manhattan where membership was by invitation only and whose German members numbered just three: Kaiser Wilhelm, his brother Prince Heinrich, and Franz von Rintelen.

But on this visit to America, as the title of his highly entertaining, self-congratulatory and selective memoir proclaims, he came forth as the 'Dark Invader', whose mission, as he stated with chilling clarity, to the delight of his masters in Berlin, was to 'buy up what I can, and blow up what I can't'. He would be terrifyingly true to his word.

On his arrival, the man who once could have lodged at any one of the city's finest clubs now booked into the anonymity of the Great Northern Hotel on West 57th Street, an establishment on the same block as Carnegie Hall, which was also used to virtuoso performances. And von Rintelen had to begin his act at once. His first battle was to win over von Papen and Boy-Ed, who received his visit to their quarters at the German Club as if he were the plague. Von Rintelen suspected that their frosty reception was really due to hurt feelings at being helped out in their covert war by a freelance operative. He warmed things up 'by informing the naval attaché [Boy-Ed] that I had been instructed to let him know that the Order of the House of Hohenzollern was waiting for him at home, and I rejoiced the heart of Captain Papen by telling him that he had been awarded the Iron Cross'.

Von Rintelen also brought a new 'most secret' German code, which he had concealed in two tiny capsules on his voyage across the Atlantic.

He handed it over to von Papen and Boy-Ed, telling them that Berlin suspected – rightly – that the British, under Blinker Hall and his Room 40 codebreaking commandos, had broken the previous German code. This new code was designed to confound the sleuths of Room 40 and provide a safe means of communication between America and Berlin.

Von Rintelen next met up with Dr Carl Bünz, the managing director of the Hamburg America shipping line, who had been working with Boy-Ed to circumvent the British mastery of the Atlantic. Bünz had moved from falsifying ships' manifests – creating fictitious cargo lists – to active sabotage, and when von Rintelen called upon him soon after his arrival in New York, the pair had an explosive conversation.

Bünz wanted detonators, and he had a plan. His men – German mercantile sailors marooned on the east coast of the US due to the British naval blockade – were running shipments of coal on chartered ships out to the open sea, to rendezvous with German cruisers for fuel transfer. They considered themselves on active duty, and partly out of patriotism, partly from boredom, wanted to do more. 'When they are sailing about on the open sea, waiting for the cruisers in order to hand over their coal, they find that time hangs heavily on their hands, so they have thought out a neat plan,' Bünz explained to von Rintelen. 'If they have detonators and meet another tramp taking shells to Europe, they will hoist the war flag, send over an armed party, bring back the crew as prisoners, and blow up the ship with its cargo. So, my dear captain, please get me some detonators.'

Von Rintelen liked the plan, but couldn't immediately see how he could go shopping for detonators without attracting the wrong kind of attention. New York's harbour had its share of pro-German supporters among the Irish dock workers, and certainly among the stranded German sailors, but that incendiary cocktail had not escaped the eyes of the Americans, nor the British, who had placed agents on the New York waterfront to watch out for saboteurs. Guy Gaunt had persuaded J. P. Morgan to hire men from Dougherty's Detective Bureau and Mercantile Police, a well-connected business run by former New York deputy police commissioner George Dougherty and his brother Harry, to patrol the docks and protect shipments from sabotage – a cost that Morgan handed off to the British. British taxpayers were financing not only their American war goods, but also the muscle to make sure they arrived intact.

Bünz connected von Rintelen with German-American Max Weiser, an exporter whose business was hamstrung by the war. Together the duo set up the import-export firm of E. V. Gibbons (the same initials as on von Rintelen's false passport) on Cedar Street, a pleasant stroll from von Papen's War Intelligence Centre in Lower Manhattan. Von Rintelen filed the company with the Commercial Register, and he and Weiser started sending out letters inviting firms to ship 'wheat, peas, shoe polish, glassware, rice, and similar goods. We posted piles of letters, so that our firm might present the appearance of a flourishing concern.'

Von Rintelen charmed and flattered the New York arms dealers, but to no avail: a sympathetic merchant of war showed him just how much the Italians were offering for the same explosives that he wanted to buy. Von Rintelen knew two depressing things: the Italians were going to join the war on the side of the Allies (they did so in May 1915), and his $500,000 wasn't going to buy him much of anything in this market.

Then luck walked in the door. Soon after setting up his shell company, von Rintelen had an unexpected visitor, sent by von Papen to provide lethal assistance. Dr Walter Theodore von Scheele had, until the outbreak of the war, been Germany's secret weapon in America. He had studied pharmacology and chemistry in Bonn, and his middle-aged face bore the scars of student duels. He had served as a lieutenant in Germany's Field Artillery Regiment 8, and in 1883 emigrated to the United States to pursue further research. His military leave came with a condition: he had to make himself available to Germany's military attaché in Washington. Scheele was assigned to track and report on American discoveries in explosives and chemicals that could be used in war. He was paid an annual retainer of $1,500, and while living the life of a mild-mannered pharmacist with a shop in Brooklyn, his industrial espionage was so good that he was never recalled for military service to Germany, and indeed, rose in rank from lieutenant to major.

When Scheele met von Rintelen, he was not only Germany's longest-serving spy in the USA, but he had the perfect cover as president of the New Jersey Agricultural Chemical Company, which he'd created in 1913 in Hoboken under orders from von Papen. And by way of welcoming von Rintelen back to the US, he reached into his pocket and placed a cigar on the Dark Invader's desk.

Except this was not a cigar. It was a lead pipe the same shape and length as a good Cuban 'torpedo', and hollow within. Scheele had inserted a circular copper disc halfway down the pipe, and soldered it in place to create two separate compartments. One would hold picric acid, an explosive; the other an inflammable liquid such as sulphuric acid. The device was an ingenious time bomb: the thickness of the copper disc determined the speed at which the acids ate through it to unite in combustion. Best of all, it was easily portable. It could be placed on munition and supply ships in the harbour by sympathetic (or bought) dockworkers, timed to burst into ship-crippling flame somewhere out at sea, and so keep suspicion out to sea as well. All von Rintelen needed was a place to manufacture the bombs.

Enter the shady waterfront lawyer Bonford Boniface, a tall, lean rogue whose 'pince-nez … kept on slipping down his nose, and gave one on the whole the impression of a mangy hyena seeking its daily prey on the battlefield'. Boniface, who always smelled faintly of whisky, knew just the man to help with the cigar bomb plan: Captain Karl von Kleist, a 70-year-old retired German naval officer who lived in Hoboken. Von Rintelen was an old family friend of von Kleist, 'a funny little old man who looked like a cartoon of the late Prussian eagle'. But von Kleist was no cartoon German. He knew the ways of the harbour, and the captains and officers of the interned German ships. And he had a brilliant idea that would help von Rintelen to create a munitions plant of his own, right under the noses of the US and British: 'We were to transplant ourselves, with all our schemes, devices, and enterprises, on board one of the German ships and thus place ourselves in a most admirable situation. Germany within American territorial waters! What possibilities!'

And so the interned ship SS *Friedrich der Grosse* became a bomb factory. Von Rintelen used the good offices of E. V. Gibbons to order bulk supplies of lead tubing and copper rods, along with the equipment to cut them. He then set up a shell-making operation on board the ship. Once the cigar tubes were cut, and the timing disc inserted, they were spirited under cover of darkness to Dr Scheele's laboratory in Hoboken to be loaded with the explosive cocktail. Soon von Rintelen's factory was making 50 cigar bombs a day.

Towards the end of April 1915, the SS *Cressington Court* caught fire in the Atlantic, while two bombs were found in the cargo of the SS *Lord Erne* and another in the hold of the SS *Devon* – all of them Allied supply ships out of New York Harbor. In May, three more supply ships either caught fire or had bombs discovered on board, and explosions rocked a DuPont powder factory in New Jersey.

In the space of three months after his arrival in America, von Rintelen's war machine was up and running, and the 'most secret' code that he had couriered to New York was still, as far as he knew, unbroken by Room 40. However, the codebreakers in Hall's diplomatic section were actually getting close to deciphering it, while his agents in the US recognised the danger von Rintelen presented and were already on his trail. Even though he was in New York under a false name, von Rintelen had cut a swathe through New York's social scene during his time as a banker, and he did not spend his evenings in Manhattan hiding from British agents. Still, he was proving difficult to pin down. And he was getting bolder.

The men from Dougherty's Detective Bureau received what seemed a gift when a German sailor with a fondness for drink was heard loudly boasting, falsely, in a tavern that he was the Captain Rintelen who put bombs on ships bearing war materiel for the Allies. In the kind of moment of farce that wars often produce, the real von Rintelen collided with this story while lying low at the seaside – and the teller of it was none other than Guy Gaunt, Hall's spymaster in America.

Though Gaunt's position was officially diplomatic, he exulted in his role as espionage chief, which theoretically at least would have resulted in his expulsion should he be discovered by the neutral United States. He made the possibility of discovery even easier with a cocktail of naval swagger and colonial snobbery spiced with insecurity, all of which fermented into a blustering ego that enjoyed the attention of New York society ladies when hints about his involvement in the dark arts were dropped at city galas and summer homes.

Franz von Rintelen, Gaunt's equal in the self-promotion department, often repaired to a hotel near Stamford, Connecticut, and while enjoying the Atlantic air he met some fetching ladies who invited him to a party at an exclusive hotel. At the party, he found himself face to face with Guy Gaunt, the man who was hunting for him at that very moment. Rintelen, with reckless bravado, introduced himself as Commander Brannon, a fellow Englishman and naval officer.

After pleasantries, Rintelen got down to finding out just what his pursuer knew. 'We have heard so much in the last few weeks about acts of sabotage against our ships,' he ventured. Guy Gaunt, showing remarkable naiveté (or perhaps Rintelen was showing a mastery of disguise or rhetoric, or both), replied, 'There is a gang working in New York Harbor under the direction of a German officer. We even know his name. He is called Rintelen, and has been mentioned a number of times in wireless messages by the German embassy ... He even admitted his identity once in a tavern, when he was drunk, and hadn't a hold on his tongue. He did not give away any details concerning his activities, but it is certain that he owns a motorboat, and runs about in it for days together selling goods of all kinds to the ships in the harbour. I cannot tell you any more, Commander, but I can promise you that he soon will be in our hands.'

Von Rintelen was elated by this news, as Gaunt's intelligence, while essentially correct, was so misplayed by its conveyor that Rintelen believed he was actually safe. But he failed to register that the British had intercepted and read his wireless messages.

While von Rintelen was devising acts of sabotage and sipping cocktails at the seaside, Emanuel Voska's team was busy trying to crack the German war machine in New York, feeding information to Gaunt, who would pass it on to Blinker Hall and Room 40. Voska had in his service clerks, messengers, waiters, maids, chauffeurs, and the assistant chief clerk in the Austrian embassy. One of his female agents, through money and charm, had convinced an employee in Karl Boy-Ed's office to steal the 'most secret' code. Von Rintelen didn't know it, but the British had him and his sabotage in their sights.

In the end, the Dark Invader was most likely brought down by a combination of von Papen's carelessness when sending cables to Berlin, openly using Rintelen's name and discussing his activities, and the extraordinary actions of a German-American. On Friday 2 July 1915, the beginning of the Independence Day long weekend, Erich Muenter planted a bomb in a place that should have been among the most secure in the land: the Senate wing of the Capitol Building in Washington DC. He had timed it to explode when the building was deserted. When the bomb detonated near a telephone switchboard, plaster was torn from the walls and ceilings,

*The destruction caused by Erich Muenter's bomb in the
Senate wing of the US Capitol Building, July 1915*

mirrors and chandeliers were shattered, doors were blown open – one of them a door into the vice president's office – and the east reception room was destroyed.

Muenter wrote a letter to a Washington newspaper protesting against munitions shipments to Germany from the USA, then took a night train to New York. There he transferred to a service taking him to Glen Cove, Long island, where J. P. Morgan Jr was breakfasting in his summer house with his esteemed guest Sir Cecil Spring Rice, British ambassador to the USA. Muenter burst in upon them brandishing a revolver and shot Morgan twice, seriously wounding him but not preventing the sturdy financier from tackling his smaller assailant and pinning him to the floor.

In jail, Muenter withdrew his alias of 'Frank Holt' and confessed to being the fugitive professor of German at Harvard University who was . wanted in connection with the murder by poison of his wife in 1906. He had acted for the Fatherland, he said. Alarm bells rang for the military intelligence strategists in Berlin. They thought that if Rintelen was behind this dangerous act of sabotage at the highest level of the US government, as well as a spectacularly public assassination attempt on America's leading banker (while in the company of the British ambassador), then he'd clearly forgotten the covert nature of his mission.

On the morning of 6 July, von Rintelen was enjoying breakfast at the New York Yacht Club when an attendant informed him there was a phone call for him. On the other end of the line was naval attaché Karl Boy-Ed, who told the Dark Invader to meet him on a street corner, where he handed over a terse telegram from HQ: von Rintelen was being recalled to Germany, effective immediately. Later that night, conveniently for everyone, Muenter walked out of his open jail cell and plunged head first to his death on a concrete floor below.

Rintelen departed for his Berlin debrief as he had arrived in New York: under the Swiss passport of Emile Gaché. On the first night of the voyage he went to the dining room and ordered a bottle of wine, to find solace in the grape. There he found that he was recognised by a German aristocrat whom he had seen often in Berlin society. The man, the Count of Limburg

Stirum, worried about von Rintelen's safe passage; ever brazen, von Rintelen assured the man that he was a Swiss diplomat and had been thus when they'd met in Germany.

For the rest of the voyage, von Rintelen dodged the count, worried that he'd eventually remember his real name and the details of their encounters. Finally, the chalk cliffs of England lay to port of the SS *Noordam*, and in the full day it took to sail past them von Rintelen 'found it necessary to visit the bar at intervals to fortify myself'. On the morning of Friday 13 August, he was interrupted in his bath by news that British officers wanted a word.

Von Rintelen made a great game of it, charming his captors and winning their sympathy. He even managed to survive the accusations of a Belgian waiter who used to work at the Hotel Bristol in Berlin, and who now recognised him during a tea break at the hotel in Ramsgate where he had been taken for questioning. Von Rintelen was released, and with happy thoughts of the Fatherland on his mind was being ferried back to the *Noordam* to resume his voyage when his luck ran out. He was recalled to land for one more interview. At Scotland Yard.

This time his interrogator was none other than Blinker Hall, who had brought with him Lord Richard Herschell, his private secretary and a key member of the Room 40 codebreaking team. At a heavy table to the left of the fireplace sat the bespectacled head of Special Branch, Sir Basil Thomson, a close friend of Hall. Von Rintelen knew that he had to play his best game yet if he hoped to evade the high-powered inquisition in front of him.

He thought he had succeeded in convincing Hall and his team that he really was a Swiss businessman, and was taken, as per his demand, to the Swiss legation for protection. While there, he heard his British minders talking about Blinker Hall's canny decision to contact the Swiss authorities in Berne to see if Emile Gaché was at home, or if he could be found in London.

Von Rintelen quickly realised that if his ruse was uncovered, the British could send him back to America as a spy, in store for undoubtedly rough justice. On a rainy August night in London, he put his German uniform back on, so to speak, and demanded an audience with Blinker Hall, where he reintroduced himself by saying, 'Captain Rintelen begs to report to you, sir, as a prisoner of war.'

Hall, appreciative of the theatrics but as ever one step ahead, congratulated von Rintelen on his subterfuge, and Lord Herschell made them all cocktails. Then the two men, in the guise of officers and gentlemen, took von Rintelen to their club for dinner before dispatching him to prison camp. It was during this dinner that the British showed just how deep their intelligence ran. As von Rintelen recalled in his memoir:

'You need not have waited so long for that cocktail I gave you at the Admiralty, Captain,' said Lord Herschell to Rintelen when they were seated in a comfortable corner of the club.

'So long?'

'We expected you four weeks ago. Our preparations had been made for your reception, but you took your time. Why did you not leave New York as soon as you got the telegram?'

Suddenly von Rintelen realised that the British had been reading German telegrams for as long as he had been in America. Just to drive home the point, Blinker Hall pulled a piece of paper from his pocket and read aloud the very message that von Rintelen had received from Boy-Ed calling him back to Germany. Indeed, the British had only lost track of von Rintelen when he had first brought the new code to New York.

'You had hardly got there when they started using it,' Hall told him, enjoying himself. 'Of course, we had been informed that you were coming, that you were going to America and taking a new code over; all that had been telegraphed to New York, and we had read it. From that moment we were unable to decipher your people's telegrams any longer, till we got hold of the new code too.'

As von Rintelen was driven off to prison camp, he was staggered that the British knew everything that was sent between the USA and Germany. But he also knew that he had left one other plan in place that hadn't been telegrammed to anyone. It would be the Dark Invader's fatal legacy.

# Chapter 7

# HOMELAND SECURITY

On the warm afternoon of 24 July 1915, the publishing propagandist George Sylvester Viereck, editor of *The Fatherland*, departed the Hamburg America Line offices in Lower Manhattan in the company of another man. The duo caught the Sixth Avenue elevated train uptown, and, busily conversing in German, didn't notice that they were being followed by Frank Burke, a wiry five-foot-seven bantamweight who was the head of the US Secret Service's special section, quietly created by Woodrow Wilson in May 1915 in the wake of the *Lusitania* sinking. Burke's brief was to ferret out spies among the millions of people in the USA with direct or distant connections to Germany and Austria – immigrants and US citizens alike.

Travelling with the 46-year-old Burke on that Saturday in July was agent W. H. Houghton, the two feds on a seemingly uneventful tail that would in fact turn out to be the first major intelligence victory by the USA in the war at home.

As Burke and Houghton followed Viereck and his companion, they couldn't help but notice the deference with which Viereck spoke to the other man, who with his trim moustache and fat briefcase looked like an accountant. It was the sabre scars on the man's cheeks, however, that were the catalyst for Burke to think hard about where he had seen that face before.

Viereck got off the train at 23rd Street and Houghton followed him. Burke stayed on the train to watch the man whom he now realised was Dr Heinrich Albert, Germany's commercial attaché to the United States. A young woman came and sat next to Albert, who moved his briefcase and lost himself in a book. Or he fell asleep. Accounts vary, with some suggesting that Viereck, acting as a double agent, had drugged Albert so that he would nod off, and that the Czech agent Emanuel Voska was actually the one who slipped into the story and snatched his briefcase. The version that Frank Burke recalled with certainty when he was 73 years old was that Heinrich Albert nearly missed his stop at 50th Street. At the last minute, he leapt up and hurried off the train. The young woman called after Albert that he had forgotten his briefcase, but Frank Burke already had it in hand, and told her to relax, it belonged to him. Then he followed Albert off the train.

By the time Albert had realised what he had left behind, Burke – who could run 100 yards in ten seconds – had sprinted down the stairs to the street and on to a trolley car. Albert, desperate and sweating, ran after him, but Burke told the trolley conductor that the German was deranged, and the conductor moved the trolley onward. Albert could only watch as the secrets of Germany's war in America rode off in the arms of the US Secret Service.

The papers in the briefcase included accounts of the pro-German stories that von Bernstorff had planted in the US newspapers. There were documents relating to how the American Correspondent Film Company had been set up to produce front-line propaganda for American audiences. There were accounts of monies paid to professors to write flattering books about Germany, and most incriminating of all, documents relating to the creation of the Bridgeport Projectile Company, a phoney arms manufacturer that would buy up munitions supplies and services solely in order to deprive the Allies of both.

The US government couldn't admit that one of its agents had stolen the briefcase of a German diplomat – an Imperial Privy Councillor, no less, who enjoyed immunity from prosecution. So President Wilson's trusted adviser and confidante, the obliging Colonel Edward House, leaked the documents to the press.

Heinrich Albert had placed newspaper ads to retrieve his lost briefcase, and on 15 August he was mortified to see that the newspapers had responded: splashed across the front page of the *New York World* was his damning correspondence. The newspaper, without revealing where it had obtained the evidence, itemised Germany's spy crimes, and named von Bernstorff, von Papen and Albert – jokingly referred to thereafter as the 'Minister Without Portfolio' – as foremost in their perpetration.

Despite all this publicity, Woodrow Wilson still played the caution card, seeing himself as the world's peacemaker, while Ambassador von Bernstorff coolly characterised the incident as some unsporting American trick and, moreover, not of any consequence: 'The affair was merely a storm in a teacup; the papers as published afforded no evidence of any action either illegal or dishonourable.' In fact, von Bernstorff was covering the fact that despite his position, he wasn't privileged to know everything that German agents were getting up to in the US – von Papen and his crew used their own set of codes to communicate with Berlin, which the British intercepted but von Bernstorff did not.

Even so, the evidence was piling up. From the time of Albert's lapse on the Sixth Avenue train to the middle of December, there were 23 acts of sabotage in the United States, or on ships sailing from her, as well as two in Canada carried out by German agents based in Detroit. Trains carrying munitions exploded, ships caught fire, munitions factories were destroyed by explosions or fires, and German agents were arrested and confessed, implicating Franz von Papen as the mastermind of their actions. The net was tightening.

On 19 August, another White Star passenger liner, the RMS *Arabic*, was torpedoed without warning by a U-boat, and sank in just nine minutes off the coast of Ireland, near the spot where the *Lusitania* had gone down three months earlier. Forty-four people died, including three Americans, and the United States was now confronting a crisis: if the Germans had sunk the *Arabic* deliberately, they would break off diplomatic relations, a prelude to war.

On the night that the *Arabic* went down, Count von Bernstorff was having dinner on the roof terrace of the Ritz-Carlton with Dr Constantin Dumba, the Austro-Hungarian ambassador to the US, and James Archibald, an American journalist. Archibald's German sympathies had landed him on the payroll to write pro-German stories in American papers, deliver pro-German lectures, and courier sensitive documents through the British blockade to Germany and Austria.

Dumba gave Archibald a packet of documents to take across the Atlantic, but the British knew he was coming because of the messages that Blinker Hall's team was now decoding with the help of Germany's no longer 'most secret' code, which had been delivered to von Bernstorff and company in April 1915. On 1 September, Archibald was hauled off the ship at Falmouth, and British agents searched for his briefcase containing the documents from Dumba. They even tore out wall panels in the saloons and lounges, but couldn't find it. There was one place left to look: the captain's safe, a sacrosanct spot staunchly defended by the ship's skipper. Blinker Hall's response to a naval protocol that he knew so well was unsentimental. He sent a squad of sailors and a couple of locksmiths on board to issue an ultimatum: if the captain didn't open his safe, they'd open it for him. The captain opened the safe.

In Archibald's briefcase, the British found documents from the Austrian ambassador and his government planning to launch strikes and labour unrest at American munitions plant, with von Papen's approval. In one letter Dumba disparaged President Wilson, and in another, from von Papen to his wife, the United States was the target: 'I always say to these idiotic Yankees that they should shut their mouths and better still be full of admiration for all that heroism.'

Despite the secret agreement between Germany and the USA that the Germans would no longer torpedo passenger liners, and German insistence that the sinking of the *Arabic* was an accident, Woodrow Wilson had finally seen enough, relatively speaking. He ordered an investigation into the activities of the German military and naval attachés in the United States, which produced more evidence of the secret war waged from within by von Papen and Boy-Ed. Wilson demanded their expulsion, and the duo were declared *personae non gratae* and recalled to Germany, despite von Bernstorff's protests of persecution: 'both [von Papen] and Captain

Boy-Ed were constantly attacked in the anti-German press, and accused of being behind every fire and every strike in any munition factory in the United States'.

It was the returning Franz von Papen that Blinker Hall now had in his sights. When his ship appeared off the coast of England, a British boarding party paid a visit. Even though von Papen was travelling under an American guarantee of safe conduct, this did not include all the incriminating documents, enciphered and not, that he carried in his steamer trunk. In his possession was a chequebook showing deposits of more than $3 million from the secret war's paymaster Heinrich Albert, as well as stubs bearing the names of suspected or already subpoenaed German saboteurs. Hall released details of his haul to the newspapers, hoping that this further proof of German treachery would bring America's warriors to the Allied side.

It was not yet to be.

Back in Berlin, von Papen and Boy-Ed were decorated and dispatched back into service, with von Papen winding up a battalion commander and seeing action at the Somme and Vimy Ridge before promotion to the general staff and transfer to Palestine. Boy-Ed resumed work with German naval intelligence. The expulsion of two major players in Germany's campaign in America would seem to have been a winning blow for the Allied cause, though the Germans were far from defeated in the USA. But now the Americans were starting to fight back.

Despite the efforts of Blinker Hall's men to help the country they hoped would soon become their ally by gathering intelligence the US could not, Germany's subversion and sabotage campaign benefited immensely from the primitive and unwieldy state of the USA's secret services. When war broke out in August 1914, the United States had no national security service it could command to ensure that it didn't become a North American front for a European conflict. As a result, it relied on the Justice Department's Bureau of Investigation (which would add the prefix 'Federal' in 1935), the Secret Service of the Department of the Treasury, and local police forces, abetted by private contractors such as the Pinkerton Detective Agency, the venerable private investigation firm formed in 1850 that had been personal bodyguard to Abraham Lincoln during the US Civil War. The Pinkertons, along with Dougherty's agency, buttressed the paltry 300 federal agents collecting and countering intelligence for the USA in 1914. As far as German sabotage and spying in the United States went, the Americans still looked at it through the lens of police work.

This meant that Captain Thomas Tunney, leader of the New York City Police Department's bomb squad, had his work cut out. Tunney, with his trim moustache and gimlet cop's eye, had joined the NYPD in 1897, when he was 24 years old. As an Irish Catholic serving in a police force with a formidable Irish Catholic presence, Tunney's intelligence, integrity and diligence – more than his tribal identity – propelled his career forward.

He became interested in bombs in 1904, when European anarchists brought their war to New York City, and caused considerable property damage. Over the next decade he would learn about the various kinds of

*Captain Thomas Tunney of the New York Police
Department Bomb and Neutrality Squad*

explosives in the bomb-maker's arsenal, their relative strength, how they were detonated, and the containers used to house them. Equally important was the handling and disposal of unexploded bombs.

Tunney was an acting captain when Police Commissioner Arthur Woods created the bomb squad in August 1914, in the aftermath of an accidental explosion on 4 July in an apartment in Harlem. That blast killed three members of an Italian anarchist group who were building a bomb destined for the Rockefeller estate in Tarrytown, New York. The squad's mandate was to investigate and suppress the activities of anarchists and the city's various organised blackmail and extortion gangs, such as the Black Hand, who used bombs as their weapons of choice when they weren't shooting, garrotting or stabbing. As the war in Europe settled into a long-term slaughter, their investigations expanded to include the activities of Germans and German-American sympathisers attempting to disrupt the flow of Allied shipments of food and war materials through New York Harbor. As Tunney put it, 'it took no superhuman amount of reasoning to combine the abnormal destruction of property in New York with the strong suspicion of German activity'.

And so the Bomb and Neutrality Squad was formed, working not in the depths of a government building solving enciphered puzzles, but pounding the pavement to decode just why one German was so busy along the New York waterfront in the autumn of 1915.

Paul Koenig, 'a sort of cross between Dr Moriarty and a gorilla, a slippery conniver one minute and a pugnacious bully the next', as Tunney described him in his lively memoir *Throttled!*, was the chief detective for the Hamburg America Line, a German steamship company that had transported generations of German emigrants to the United States. Koenig, derisively called 'the bullheaded Westphalian' by workers behind his brawny back, had patrolled the New York/New Jersey waterfront for Hamburg America since 1912. His duties had included investigating fires, theft, stowaways and charges against officers on the line's ships. Then, in late August 1914, his diary recorded that he had begun work for a higher power, the Fatherland: 'Aug. 22. German government ... entrusted me with handling a certain investigation. Military attaché von Papen called at my office later and explained the nature of the work expected. (Beginning of Bureau's services for Imperial German government.)'

In September, Koenig met von Papen and was instructed to set up a 'Bureau of Investigation' for him. This he did with ruthless efficiency, and by the time he caught Tunney's attention, his network had thoroughly penetrated the New York docks. Tunney couldn't help but notice that Koenig was 'curiously busy' at a time when the German ships he watched over were lying idle, and that in turn made Tunney and his men 'busily curious to find out why'.

Their initial attempts to shadow the huge and shrewd Koenig proved amusing to their target, who, as a detective himself, knew the tricks of tailing, and would pop out of the shadows to laugh in the faces of the New York cops who thought he hadn't noticed them following him on his missions. Tunney and his men – a ragout of ethnicities that reflected the melting pot of America, and would prove useful to undercover work –

FROM THOMAS J. TUNNEY, *THROTTLED!*

*Paul Koenig, chief detective for the Hamburg America Line*

adapted their methods, and with a more subtle form of 'team tailing', where one man would hand off surveillance to another, they were soon able to keep discreet tabs on Koenig.

Koenig's ambit in New York was intriguing: from the Hamburg America offices in Lower Manhattan, he would make regular visits uptown to Pabst's German restaurant in Columbus Circle, to the German Club on Central Park West, to Lüchnow's in Union Square, as well as to the Belmont and Manhattan hotels, which Tunney lamented had direct connections to the subway from their basements, and provided saboteurs and murderers alike with quick and effective escape routes.

Tunney decided to 'cut in on the wire' of the Hamburg America office phone, and finally caught a break: one of Koenig's operatives was furious at the intimidating detective over that age-old source of conflict, money. The call was traced to a pay phone in a saloon, and a bartender with a good memory: the man they were looking for lived around the corner. Police work did the rest, and so George Fuchs, the man with the grievance against Koenig, was invited to come for a job interview at a phoney German wireless company set up by the NYPD.

Detective Valentine Corell, who spoke German, played the part of the office manager, and after establishing that Fuchs was easily the best man for the job took him out for a drink at a nearby bar. Over mugs of beer, he teased out the story of Fuchs' grievance. Fuchs had done a lot of espionage work for Koenig, even scouting out the Welland Canal and working on a plan to blow it up. As an employee of Koenig's Bureau of Investigation, at $18 a week, he spied on ships and cargo leaving New York, did special guard duty at Heinrich Albert's office, and acted as general muscle for German dignitaries. He had called in ill on a recent Sunday, and Koenig had responded that 'illness should never interfere with service to the Fatherland'. He had fired Fuchs, citing his 'constant quarrelling ... drinking, and disorderly habits', and then made his fatal mistake: he refused to pay Fuchs the $2.57 due to him in back pay.

Tunney now had enough evidence to arrest Koenig, but as the crime was federal, he turned the evidence over to the Department of Justice. With

Koenig in custody, Tunney and the detectives who had broken open the case were amazed by his little black memorandum book, which 'told the story of the Bureau of Investigation with a devotion to detail almost religious'.

It also told that story in code. Koenig had devised a simple but effective system for keeping track of his activities, with aliases for himself, and code numbers for his contacts: the German embassy was 5000; von Papen 7000; Boy-Ed 8000, and so on. He also changed code names frequently for rendezvous spots – 'Pennsylvania Station' was code for Grand Central Station at one point, and 'Brooklyn Bridge' rather fancifully meant the bar in the Unter den Linden restaurant. If a meeting was to take place in Manhattan, the address as revealed over the telephone would be five blocks further than the actual street number (i.e. a meeting supposedly on 15th Street would really be on 20th).

Koenig changed these codes weekly, and his black book provided ample evidence to the Bomb and Neutrality Squad of Germany's distinctly non-neutral actions in the USA. Koenig was eventually imprisoned for his role in the Welland Canal plot, but his capture did not stop the Germans from waging their war in America. The network of German agents was now diverse and determined to build on the success it had seen, and the astonishingly light consequences for its actions meant that now Germany would be even bolder in attacking targets in the United States.

Blinker Hall and Room 40 had become adept at intercepting and decoding cables from Berlin to the embassy in Washington, and their intelligence, said Maurice Hankey, Secretary to the Committee of Imperial Defence, was priceless as it gave the British war planners information about how Berlin was viewing the war in Europe by virtue of how widely and deeply it wanted to wage the war in the United States. Despite the activities of German agents in the USA, an intercepted communiqué between German chancellor Bethmann-Hollweg and von Bernstorff revealed that the Germans saw the USA as a welcome peace-broker, if Wilson committed to a peace plan soon, before England could get the upper hand.

But the British wanted more than intelligence; they wanted the USA to realise that it was already at war with Germany, and to join in a fight that had now spread around the world. The Turks had attacked the British at the Suez Canal and begun their genocide of the Armenians; Italy had entered the war on the side of the Allies, and Bulgaria on the side of the Germans; in Africa, the Allies captured Namibia, while the Germans conquered Warsaw; and the Allies attacked Turkey via the Gallipoli Peninsula, suffering a quarter of a million casualties and a defeat so severe that Winston Churchill, First Lord of the Admiralty and champion of the doomed battle plan, was consequently demoted before resigning in humiliation. He then joined the battle on the Western Front as a lieutenant-colonel commanding the 6th Batttalion of the Royal Scots Fusiliers.

The Western Front was an ever-expanding mass graveyard. The Germans had used poison gas for the first time in history at the Second Battle of Ypres in April 1915, which the Allies held at the cost of nearly 70,000 casualties; the Allies reciprocated by gassing the Germans in the futile Loos offensive in September and October, which saw another 50,000 British casualties, as well as the sacking of their commander, Sir John French, who was replaced by Sir Douglas Haig. And despite the sinking of the *Lusitania* and German promises to respect neutrality, German U-boats continued to kill Americans, until US protests led them to call off unrestricted submarine warfare in the Atlantic in September 1915. The Germans simply redeployed their U-boats to attack shipping in the Mediterranean. In more than a year of the most epic combat the world had ever seen, all that had been decided was that things were getting worse.

In a New Year's Day 1916 editorial, *The New York Times* presented the cost as rendered at the Pan-American conference then meeting in Washington, one whose bureaucratic cool failed to mask the galloping expense: 'The money cost of the war to July 31 this year, exclusive of the capitalized value of human life, was estimated at $37,696,774.00 … By January 1, the aggregate would be $55,000,000.00, and should the war continue at the end of the second year, next August, it would reach $88,000,000.00. At the end of the second year the probable human loss was estimated to be 12,000,000 lives. The capitalized value of those lost workers was placed at $35,196,000,000.00.' The piece also reported that the USA had loaned a dozen of the hostile nations more than $889,000,000 at a yield of 'six per cent or more ...' Adjusted for inflation, as US dollar in 1916 was worth 22 times what it is today. The USA had, within less than two years, become banker to the world.

In London, the intelligence powers knew the human and financial cost all too well, and decided that it was time to give their American cousins some more help in fighting the enemy within, if only to fuel the Allied cause by preserving 'guns and money' from the USA until the Americans joined the Allied fight to help win it as soon as possible. On 28 October 1915, two Englishmen had arrived in New York City seemingly to do business with Touche and Niven, an accountancy firm in Lower Manhattan connected to the parent firm of George A. Touche & Co. of London. The older man, Sydney Mansfield, was about 40, with distinguished grey hair and the manner of an affluent businessman – which he really was, with the private British investment firm of Hendens Trust. His younger companion was the 30-year-old Sir William Wiseman, who, though Mansfield's boss at Hendens Trust, was in New York on a very different kind of private investment: he was also Captain William Wiseman, a wounded warrior now on a crucial mission from his army boss, Commander Mansfield Smith-Cumming ('C'), head of Britain's Secret Intelligence Service (which would become MI6).

Wiseman had joined the army in 1914 and served with the Duke of Cornwall's Light Infantry, seeing bloody action on the Western Front. He was at the Second Battle of Ypres in 1915 – at which the Germans first shocked the Allies and the world with their use of chlorine gas – and was temporarily blinded. While recuperating back in London he had a chance encounter with 'C' – aka Mansfield Smith-Cumming – who had served in the Royal Navy with Wiseman's father. And so, in that effectively improvisational way that coloured so much of intelligence recruitment during the war, Wiseman ended up an officer in what would eventually become MI6, assigned to expand the brand to American shores.

The British were already represented in the USA by Captain Guy Gaunt, who acted as Hall's man on the ground. This brash and snobbish Australian, who revelled in his role as swaggering spy (and who was known by both Germany and the US to have extracurricular espionage duties), was not happy to receive the competition, which was how he viewed these two arrivals from England who showed up at his New York office. In the end, Wiseman would more than compete; he would quietly take over Gaunt's role in the US.

Wiseman and Mansfield would eventually return to London before the end of 1915, with Mansfield deployed to a counter-intelligence assignment in the eastern Mediterranean. Blinker Hall and 'C' decided that Wiseman would go back to America, and supplied him with the services of another wounded veteran, Norman Thwaites. British by birth, Thwaites had been educated in Germany and could speak the language fluently. He'd first gone to New York at the beginning of the century, working as a journalist for the *New York World*, and making many useful contacts in the city, the most powerful of whom was his proprietor, Joseph Pulitzer. However, when the war began, he abandoned his burgeoning career and joined up. His period of service as a cavalry officer was brief: in November 1914 he was severely wounded in the First Battle of Ypres which saw more than 200,000 total casualties over two months of fighting this new kind of industrialised war.

After recuperating from a bullet that ricocheted off a trench and through his throat and chin, Thwaites sailed into New York Harbor on New Year's Day 1916, and quickly renewed acquaintance with friends who had grown influential in his absence. His former newsroom colleague Frank Cobb was now the go-to foreign affairs man at the *New York World*; Frank Polk was now counsellor to Secretary of State Robert Lansing, effectively making him Lansing's lieutenant and the man whose office was responsible for intelligence; his pal Tom Tunney was now head of the NYPD Bomb and Neutrality Squad, chasing down German saboteurs; and his friend Charlie Dillingham, a promoter on Broadway, would end up as second-in-command of the US army's military intelligence bureau in New York in 1917.

Wiseman followed Thwaites to New York later in January, and the duo set up shop in the British consulate at 44 Whitehall Street, in Lower Manhattan, overlooking the harbour – and just across the street from the offices of the Hamburg America Line. They had recruits sent to them from England,

and they brilliantly used the resources at hand, establishing a network that included characters as diverse as occultist and provocateur Aleister Crowley, the 'Beast of the Apocalypse', and the Russian-born Shlomo Rosenblum, aka Sidney Reilly, the self-styled 'ace of spies', as well as Germans, Irish, Americans, Indians, Canadians, Mexicans and eastern Europeans. Wiseman even succeeded in stealing away from Guy Gaunt the powerful network of agents run by Czech-American spymaster Emanuel Voska.

The pair enjoyed the pleasures of New York, but never let an opportunity pass to gain the upper hand on the Germans. While dining at the Long Island home of Oscar Lewisohn, heir to a copper fortune and a social season mainstay in both England and the US, Thwaites was entertained by pictures of a recent trip that Lewisohn had taken to the Adirondack Mountains. He noticed that one of the photos in the album featured none other than Count Johann von Bernstorff, the German ambassador, with two comely young women, both in swimming costumes and neither of them the count's wife.

Thwaites managed to sneak the photo out of the mansion, have a copy made and return it to the album. His stealth attack came to light when he (with Wiseman's hearty consent) released the photo to the press, causing substantial ridicule to von Bernstorff, and great pleasure to his enemies.

In March 1916, Wiseman and Thwaites welcomed 47-year-old Robert Nathan to the fold, the third member of the trinity that would help to wage Britain's war on the North American front so effectively. Nathan had served in the Indian Civil Service, and had notably been the police commissioner in Dhaka, successfully suppressing anti-British and armed Bengali revolutionaries. He began to work for British intelligence against Indian opposition groups in 1914, and after returning to England served on the Western Front as an interpreter for Indian troops, the perfect cover for an intelligence agent looking to overhear Indian plots against the state. His brief in New York was still to pursue Indian sedition.

'C' and Blinker Hall now had in place three agents to fight their intelligence war in North America: to investigate suspects about whom the home office needed information; to keep track of the Irish Nationalist movement in the United States; and to probe the murky and dangerous world of 'Hindu sedition'. Nevertheless, despite the trio's presence, the departed von Papen and Boy-Ed had created an extensive network in America, and had left a large and committed force of German agents behind. Germany was fighting a war on three fronts, and their biggest attack on America was soon to come.

# Chapter 8

# DEATH FROM THE SKIES

In early 1915, Kaiser Wilhelm, a man better known for his bellicose rhetoric than his humanitarianism, was on the horns of an ethical dilemma. Should he bow to the urgings of his military and allow the Zeppelin fleet free rein to attack Britain regardless of the civilian casualties that would inevitably follow? Or should he exercise restraint? A bombing campaign, whatever its material successes, could easily backfire in the court of international public opinion, putting Germany beyond the pale in the eyes of neutral states, particularly America.

As the pressure on him mounted, the Kaiser equivocated: on 9 January he gave permission for raids on Britain but not London; on 12 February he admitted the London Docks to the list; on 5 April he extended it to include the East End but warned his commanders not to hit residential areas or the royal palaces – after all, members of his family lived there. Then, on 20 July, he agreed to unrestricted bombing. At last the advocates of the Zeppelin had the green light to pursue their vision.

The airship was named after the man who had created it, Count Ferdinand von Zeppelin, a member of the Engineering Corps who was so impressed by the observation balloons he'd witnessed first hand during the American Civil War that he decided to design his own version. With a rigid framework 420 feet long and 38 feet wide, 4,000 cubic feet of hydrogen to keep it afloat and two 16 h.p. motors to run it, the first model, LZ1, was test-flown in 1900.

Both the army and the navy seized on this new weapon, each developing their own version with rival manufacturers – the army ships were designated

S-class, the navy L-class. By 1914, there were 11 Zeppelins, armed with crude incendiary bombs and explosives, with more coming off the production line and teams of technicians working to create ever bigger, faster and more durable models.

The first foray was mounted by Zeppelins L3 and L4 on 19 January 1915. Eleven incendiary bombs were randomly dropped on the seaside town of Great Yarmouth: one hit a church, killing two people, including a 72-year-old spinster; a shoemaker was killed in his shop; and one bomb hit a house where a young girl was playing the piano: 'the force of the explosion lifted her completely from the piano stool and planted her a distance away'. Thankfully, she survived with only minor cuts and bruises. Meanwhile, 16 bombs fell on nearby Hunstanton, leaving two dead.

For the British people, the experience of seeing these immense ships appearing in the sky above them must have felt like that moment in sci-fi movies when the alien spaceship first emerges from the clouds. Though H. G. Wells had anticipated their arrival in his visionary *The War in the Air* (1908), nothing could have prepared the average citizen for the reality of death on their doorsteps. One civilian remembered how his 'old pal' reacted to seeing them: 'he was half scared out of his wits. He was breathing heavy' and thought they resembled 'big silver cigars'.

After a break while the Kaiser vacillated, raids resumed mid April. There were five more in the next four weeks, causing six fatalities as Zeppelins hit targets in East Anglia and the south-east. Then, on the night of 31 May/1 June, with the Kaiser's blessing, the first assault on London occurred. Flying in from north of the city, a lone Zeppelin hit Stoke Newington before passing over Shoreditch, Hoxton, Spitalfields and Commercial Road, swinging out again via Leytonstone, having delivered 91 incendiaries and 28 explosives, setting 41 fires and killing seven people. The first Blitz had well and truly begun.

From 1909 onwards, British attachés in Germany acting as spies for naval intelligence were reporting back on the progress of this new weapon. Two of them were treated to a test flight, and were suitably impressed by the Zeppelin's speed and manoeuvrability. In their opinion, the German authorities were contemplating using Zeppelins to drop explosives on ships, towns and dockyards.

They continued informing London about this aerial threat right up to the outbreak of war. However, it was only then, after Antwerp had been hit by Zeppelins in the first few weeks of the war, that the military woke up to the danger and began hasty preparations for home defence. As Churchill put it, Britain was 'quite powerless to prevent an attack'. Alarmed, he intervened to get things moving, thereby ensuring that the Admiralty would take charge of any counter-measures, overriding the bitter objections of the War Office, who felt that their air arm, the Royal Flying Corps (RFC), should bear the burden.

A motley collection of 68 guns was assembled in London, along with 53 searchlights, all manned by the Royal Naval Volunteer Reserve. On 1 October 1915, blackouts were introduced. By mid 1915, 60 planes, mostly from the Royal Naval Air Service (RNAS), armed with rifles and hand-held bombs, were operating in the London-Sheerness-Dover triangle, with 20 allocated to the capital.

As time went on, the resources dedicated to this defensive infrastructure increased steadily – by the end of 1916, 17,000 servicemen were watching the skies, adding to the dangers facing the Zeppelins. A veteran German captain compared the searchlights to 'loathsome rigid snakes that have eyes', holding their victims in 'electric chains'. Once illuminated, bombardment followed: 'Hell's orchestra breaks loose … sirens howl, mortars bellow, cannons thunder.'

It takes a particular kind of person to be able to sit thousands of feet in the air in a gigantic balloon filled with highly inflammable gas that other people are doing their best to ignite. Each Zeppelin carried between 16 and 23 crew – all volunteers – who had undergone six months of intensive training. They were part of the heroic elite of the German armed forces, ranked next to fighter aces and the submarine crews. Successful commanders were household names.

Travelling at such high altitudes, they had to endure bitter temperatures: 'even in midsummer our thermometers recorded 25–30 degrees below zero as soon as we were 5,000 metres up'. One officer recalled a particularly raw journey: 'it was so cold that our sandwiches had frozen in our pockets and our eyebrows had turned white with frost'. Bad weather caused constant navigational difficulties. Numerous raids were abandoned after ships completely lost their bearings.

To keep on course, they relied on communications from their meteorological and direction-finding stations. According to one veteran commander, this made the wireless operator, perched alone in his narrow cabin that was erected just in front of the engine, the key member of his crew: 'our fate depended on him keeping a cool head'.

Operators sent their first messages as the Zeppelins departed their bases. MI1(b), which was monitoring this traffic alongside Room 40, observed that 'the practice prior to a raid was for the flagship at Wilhelmshaven to hold a sort of wireless roll-call'. A signal was 'sent to the airship squadron and … acknowledged by each airship in turn'. Once en route, they relied on DF stations at Tondern, List, Nordholz and Bruges to get their bearings and negotiate the journey, especially when the ship was above the clouds or in a fog; and then, as they reached the English coast, to guide them in. This caused wireless pile-ups as each raider simultaneously tried to get confirmation of its position: 'every ship is sparking away wildly … determined to get its own information at all costs'.

All this activity was a gift to Britain's string of coastal DF stations. After the war, the art of establishing the position of a Zeppelin from its signals was discussed by H. J. Round, the theoretical and practical brains behind DF technology. When three or more stations were in action, one would act as a censor, plotting the bearings coming in from the others until they all intersected at one point. Only then was the information deemed reliable.

The Germans knew that their signals were vulnerable, 'for every electrical discharge called the enemy's attention to the fact that airships were approaching'. What they didn't realise was that their messages were being intercepted and decoded. The Zeppelins relied on the HVB code book, the same one used by the merchant navy and the submarine fleet that had been in the hands of Room 40 from early in the war. As a result, decoding their communications was relatively straightforward. It helped that the airships stuck to their call signs, making them easy to identify. Though they changed their cipher key daily, this rarely troubled the experienced Room 40 codebreakers on duty during 'Zeppelin nights'.

Hugh Cleland Hoy, who was part of Hall's staff for most of the war, remembered the feeling of anticipation as the codebreakers waited for an attack to come. They were 'continually expectant of some intercepted message from the East Coast listening stations that might give a clue to the intention of the air raiders. On January 19th, an enemy wireless communication reached them which when decoded signified that the first effort was to be made ... I still recall the breathless sensation with which I learned ... that Yarmouth had been raided.'

*The aftermath of a Zeppelin attack on Bartholomew Close, London, 1915*

On 14 June, Hoy wrote that 'decoded messages showed that L10 had left her shed at Nordholz at the same time as SL3, but there was no certainty as to her destination'. They were heading for Hull and other targets in the north-east. August was relatively quiet; then, over successive nights in early September, the capital got the worst of it: 40 dead, more than 100 injured, as both the East End and central London were blasted. A bus was flattened near Bloomsbury, while at Holborn a 660lb bomb created an eight-foot-deep hole in the ground.

The raids were getting bigger. According to Hoy, shortly after 5 p.m. on 13 October, Room 40 decoded a wireless message which showed a serious raid was pending. Half a dozen Zeppelins were involved and once again London was the primary target. Casualties were high: Hoy called it 'a fearful night' during which 71 people were killed and over 120 so badly hurt that 'probably few of them will survive to tell the tale'.

By early 1916, the Zeppelins were travelling even further afield. Room 40 decoded wireless intercepts that served as a warning of a serious attack on Liverpool. Due to bad weather, they didn't make it that far but still managed to offload 376 bombs, mostly incendiaries, on Manchester, Scunthorpe, Hull, and Walsall, where a stunned witness described the terrifying moment a tram full of passengers was hit: 'the blast broke its windows. A piece of glass struck the Mayoress who was riding in the tram and she died from her injuries.'

Throughout this onslaught the army had continued to grapple with the navy for control of home defence until a compromise was eventually reached on 10 February 1916. The Admiralty was to focus on the Zeppelins while they were over the sea, the army while they were over the UK. Room 40's intelligence remained vital, as it alerted the seaplanes of the RNAS, giving them the chance to hunt down the Zeppelins before they reached the mainland.

Hoy described how 'we learned from the German wireless that another raid was to take place. This was decoded so quickly that by 3 p.m. we were able to inform the Eastern Naval Command that there were 12 airships over the North Sea.' The RNAS sprang into action: 'our seaplanes ... were at once ordered out from Harwich ... L13 reported by wireless at about 10 o'clock that she was hit.'

Meanwhile, at the War Office, Malcolm Hay already had four officers on wireless duties. In March, they left MI1(b) and formed MI1(e) to deal exclusively with the Zeppelin menace. Two of them collated the messages that were received by the special aerial on the roof of the building within 60 seconds of transmission, while the other two attacked the daily cipher key. The unit was connected by direct line to the WO telegraph section. It forwarded the data by pneumatic tube to Room 417, the main control centre. From there the information was called through to GHQ Home Forces. One squadron leader described the scene. A large map was laid out on a table. Sitting round it were 'a number of operators ... receiving information on telephone handsets as it came in' and plotting it on the map, making it 'possible for the GOC to see the position of the raid at a moment's glance'.

With the DF stations and the codebreakers leading the way, the British had at their disposal a formidable intelligence-gathering system that produced a clear picture of the scale and timing of each raid. However, all this information was worthless unless the air force could actually destroy the enemy. Due to inadequate firepower, it couldn't. The planes simply didn't have ammunition powerful enough to penetrate the monster's skin. They were further handicapped by lack of altitude. Unable to operate higher than 13,000 feet, they were no match for Zeppelins that were still comfortable at 18,000 feet. Speed was an issue as well. The airships could climb 1,000 feet a minute; British aircraft took ten times as long to cover the same distance.

As a consequence, the Zeppelins continued their grisly work without loss. On 5 March 1916, Hull's waterfront and docks were attacked. The damage was considerable and the commander of the raid 'saw whole structures toppling into ruins until that section took on the appearance of an immense black crater on the snowy landscape'. Like other targets in the north, Hull was practically defenceless: no guns or planes were available to meet the threat. Afterwards, an angry mob stoned an RFC vehicle and attacked one of its officers. A few weeks later, a terrible tragedy unfolded in Cleethorpes when a bomb landed on a Baptist chapel full of soldiers, leaving 32 dead and 48 injured; a member of the medical team that arrived on the scene likened it to Dante's 'Inferno'.

During July, poor weather prevented a series of raids; then, on the night of 2–3 September 1916, came the biggest one of the war so far, involving 16 airships that ranged across the country, targeting Nottinghamshire, Lincolnshire, Hertfordshire, Sussex and London. However, what might have been another deeply dispiriting few hours of indiscriminate, unopposed carnage proved to be the night the tables turned.

The technological breakthrough that made the difference was the work of a New Zealand engineer, and came in the form of an explosive soft-nosed bullet loaded with phosphorus that flattened on impact, creating a large hole. Used in combination with incendiary bullets and tracer bullets, known as 'sparklets', they finally gave the air force the means to cause significant damage.

During the Saturday night in question, 2/3 September, SL11 was shot down at Cuffley Hill in Hertfordshire. News that one of the beasts had been slain was greeted by a wave of euphoria, a great release of accumulated stress and tension. The next day became 'Zeppelin Sunday'. Special trains were laid on from King's Cross to ferry curious and jubilant crowds to the crash site. The landlord of the local pub charged twopence entry.

One of the first visitors to inspect the wreckage was Blinker Hall. In May, the Germans had changed their code books. The HVB, used by the Zeppelins, was replaced by the *Allgemeines Funkpruchbuch* (AFB). For the first time, Room 40 was operating in the dark. Hall was desperate to get his hands on a copy of the new codes, especially as they were being adopted by the enemy's submarines as well. Until he did – and there was no guarantee that he would – Room 40 had no alternative but to begin the painstaking and laborious business of trying to reconstruct the AFB codes from scratch by trawling through the plethora of available messages for clues.

The obvious choice for this task was Dilly Knox, now firmly established as Room 40's leading codebreaker. By early 1916, Dilly had his own office, Room 53, a tiny space dominated by a huge desk and a bath that he had had specially installed. Lying in a steaming tub for hours helped stimulate Dilly's thought process as he pondered codebreaking conundrums. Though his work was often presented 'in inky scribbles on sheets of dirty paper', sometimes sopping wet from bathwater, there were no doubts about his ability.

With nothing to go on except messages sent in the new code, which consisted of three-letter groups, Dilly, searching for an identifiable pattern, noticed that the word ending 'en' was repeated throughout in a similar position. He realised that this construction resembled a poetic metre known as dactyls. This suggested to him that the wireless operator responsible had encoded his signature, which identified him as the sender, by using part of a poem.

Dilly was no expert on German literature, so he took the messages to someone who was: Professor Leonard Willoughby. Willoughby examined them and decided they came from a poem by Schiller. He gave Dilly the full translation of the relevant lines, and armed with these Dilly was able to identify ten new code words, which he used as the basis for further reconstruction of the AFB. Nevertheless, it might take months for him to crack the whole thing open.

The best hope of resolving the crisis, therefore, was for Hall to recover a copy of the code book from a downed Zeppelin. When SL11 met its doom on 3 September, he was ready to act fast. According to Hoy, 'Hall rushed to the scene as soon as the news of the fall of the blazing Zeppelin reached him, but by the time he reached Cuffley, the wreckage had burned itself out.'

Then, three weeks later, during another massive assault, L33 was hit several times after terrorising the East End and landed in a field near Chelmsford. Miraculously, both ship and crew survived intact. Having destroyed their code books, the men wandered to a village and were taken in by local police. That same night, 23–24 September, L32 was shot down at Snail's Hill Farm near Billericay in Essex. A fellow Zeppelin commander described the moment it was hit: 'suddenly a red light shone out vividly through the darkness … a vast ball of fire hung in the heavens … and then like a gigantic torch, the ship dropped faster and faster to earth. She crashed to the ground and continued to burn … our grief was overwhelmed by horror.'

Trooper Charles Williams was one of those assigned to guard the wreck, which 'burnt well into the early hours of the morning'. The roads leading to it were jammed with cars and people keen to get a look. By dawn, thousands had gathered. Williams and his fellow soldiers made a tidy profit selling bits and pieces of the wreckage, while local blacksmiths converted scraps of metal into bespoke souvenirs.

Hall was also on the hunt for mementoes from the remains of L32. He had a team on site before the other scavengers arrived and they were rewarded for their efforts. The new code book, somewhat charred, was retrieved from the smouldering remains.

From that point on, the law of diminishing returns applied. In the next six months, four more Zeppelins went down in flames. A further seven crashed in the sea. Not a single crew member survived. Another Zeppelin was blown up by RNAS planes during a bombing raid on its sheds at Trondern: this attack was the first ever air strike launched from a ship, HMS *Furious*. The aircraft carrier was born.

Though new Zeppelins were being built, the loss of so many highly experienced personnel was much harder to bear. The final blow was a self-inflicted wound. On 18 January 1918, a terrible fire broke out at Ahlhorn, the main Zeppelin base. Five Zeppelins were destroyed. An assistant mechanic recalled the destruction: 'I heard the benzine tanks explode, the flames crackle, the girders break and the glass burst.' Next morning, he surveyed the damage: 'I saw the lean iron ribs of several sheds standing out like leafless trees in the wintry sky. Ahlhorn had simply been wiped off the earth.'

The last attack on London was on 19 October 1917; only one Zeppelin, L45, got through. The Midlands, the north and East Anglia suffered four attacks during 1918. Wigan bore the brunt of a five-Zeppelin assault on 12–13 April. The local paper reported the death of a woman in her bed: 'a fragment of the bomb or splinter shot across the room, cut away her face and killed her instantly'. The final attempt was made on 5 August 1918. Returning empty-handed due to bad weather, Paul Strasser, the leader of the naval Zeppelin fleet, and his ship, L70, were consigned to a watery grave.

Over the course of the campaign, there had been nearly 80 raids – 26 of them on London – causing 557 deaths. Had the German incendiaries been more efficient, the toll would have been much higher. As it was, Britain was faced with a new menace, the Gotha bombers, conventional planes that threatened even more havoc. Unfortunately, the codebreakers were unable to predict their arrival, as they only used wireless on take-off, thereafter maintaining radio silence.

The first major assault came on 4 July 1917. The authorities reacted quickly, setting up a body called the Air Organisation and Home Defence Against Air Raids under the leadership of the dynamic South African Jan Smuts. It introduced a 45-minute early-warning system, linked to 80 fire stations within a 20-mile radius of Charing Cross. During eight days in late September, known as the Blitz of the Harvest Moon, the Gothas dropped 5,000 kg of bombs, killing 69 people. Nearly one million Londoners sheltered in the Underground. Others slept in parks and fields or simply left the capital. But the air force's improved armaments inflicted mounting losses on the Gothas, which at this late stage in the war the Germans could not afford to sustain, and the raids slowly petered out. Overall, through trial and error, invaluable lessons had been learned that could be applied next time Britain faced death from the skies.

# PART II

Chapter 9

# CORRIDORS OF POWER

In the early 1930s, Blinker Hall, with the help of a ghost writer, embarked on his autobiography. In the few chapters that survive from this unpublished work – it was killed stone dead by the authorities – Hall claimed that under his guidance, naval intelligence 'grew … into an almost worldwide organisation with a multitude of the most diverse activities'. This was no idle boast. His power and influence spread far and wide; nothing was out of bounds. Any action he thought would help bring victory was legitimate. He didn't care about ignoring the chain of command or stepping on other people's toes. Nor did he mind if his actions were morally or legally dubious, or carried considerable risks; as he put it, 'I came to the decision that if we were to get on with our job, there must be no slavish regard for peacetime precedents.'

Almost immediately he proved true to his word. The issue he addressed head on was that of postal interception: the opening of suspect mail. In 1911, when Churchill was Home Secretary, the system was very limited: a warrant had to be obtained for each single letter or telegram that the security services wanted a peek at. Churchill adjusted the law so that warrants covered all the mail received by suspicious individuals.

Nevertheless, in autumn 1914, there was only one censor at London's Mount Pleasant sorting office, and a few overwhelmed staff. One clerk admitted that only 5 per cent of the post was being dealt with, the rest piling up in cupboards. Once Hall was made aware of the situation, he felt 'it was imperative to enlarge very considerably the existing scope of censorship and to press for its rapid extension'.

He paid a visit to the man in charge and told him he wanted 'to make sure that *all* the foreign mails are opened and that no secret message gets through'. Without bothering to ask for permission, he got hold of 200 men, all volunteers from the National Service League, squeezed £1,600 from Churchill on false pretences, and put these resources to work on a two-month trial basis. He quickly realised that the Germans were ordering contraband goods via the post, thereby defying the blockade.

Hall's meddling nearly blew up in his face. After a few weeks, a trouble-some MP who was already under scrutiny discovered his mail was being interfered with. Outraged, he complained to Reginald McKenna, the Home Secretary, who demanded to see Hall, warning him that the sentence for tampering with His Majesty's mail was two years in prison.

With his career hanging by a thread, Hall was sent to Asquith, the prime minister, at 10 Downing Street. Though by the end of the war he had met a great number of politicians and 'seen to some extent how their minds worked', it was his first visit to this august address and he felt distinctly nervous: 'once inside this rather shabby old house I could not rid myself of the idea that I was only witnessing some strange kind of stage play. Things had suddenly become unreal.' Luckily for him, Asquith was in favour and endorsed his actions. By the end of the war, 4,801 staff, mostly women, were engaged in steaming open hundreds of letters a day.

Hall not only interfered on the domestic front, he was equally willing to take the iniative over international diplomacy. In early 1915, he put into motion an audacious plan to bribe the Turks to sue for peace. The Ottomans' entry into the war, November 1914, was half-hearted at first, reflecting deep splits in the government over the right course to take. With British warships gathering in the Adriatic, reports were coming in of the uncertainty and fear gripping the population in Constantinople (Istanbul), and Hall spied a chance to exploit these tensions.

He contacted a British civil engineer working for a firm of contractors, who was 'on friendly terms with many of the most influential Turks'. Hall's contact agreed to try and get in touch with Talaat Bey, the Minister of the Interior, who was known to have doubts about the war. In his memoirs, Hall stated that on 5 March 1915, 'the price offered for the complete surrender of the Dardanelles with the removal of all mines was £500,000'.

While Hall rolled the dice, Churchill was also prepared to gamble that the Turks would fold if enough pressure was applied. During February 1915, Churchill bullied and cajoled his colleagues into agreeing to try and send ships through the heavily defended Dardanelles and on to Constantinople; once there, the threat of the city being bombarded by their guns might prove sufficient to force the Turks to capitulate. However, the straits were guarded by fortifications armed with heavy artillery and mobile gunnery with enough firepower to prevent easy access and stall the operation.

Then Room 40 received and decoded two messages that referred to the fact that the Turkish defences, which had so far prevented British ships from getting through, were very low on ammunition and it would take several weeks before more would arrive. Could this be the chance the British were waiting for? Realising that this information might be a game-changer, Hall visited Churchill and Lord Fisher, the Admiralty's most senior naval figure, and told them the news. Both of them were galvanised by what they heard; the navy might now be able to enter the Dardanelles without sustaining heavy losses. According to Hall, Fisher exclaimed, 'By God ... I'll go through tomorrow,' while Churchill 'seized hold of the telegram and read it through again' with great enthusiasm.

Sensing that this was a decisive moment, Hall decided to tell them about the bribe, or as he put it, 'the large sum of money I had personally guaranteed'. An astonished Churchill asked how much, and Hall replied, 'Three million pounds ... with power to go to four million if necessary.' Churchill demanded to know who had authorised this offer; Hall responded, 'I did.' Alarmed, Churchill probed further: what about the Cabinet? Did it know? Unruffled, Hall replied, 'No, it does not. But if we were to get peace ... I imagine they'd be glad enough to pay.'

Before Churchill could fire off another question, Fisher, still processing the fact that the enemy was running short of ammunition, declared, 'No, no, I tell you, I'm going through tomorrow, or as soon as can be completed,' then turned to Hall: 'Cable at once to stop all negotiations … We're going through.'

Hall bowed to his superiors and took the bribe off the table. The navy renewed its doomed assault on the straits, frustrated as much by the dense network of minefields as by the Turkish guns. After repeated attempts the operation was abandoned and the fateful decision made to conduct an amphibious landing at Gallipoli. The troops involved quickly became mired in trench warfare every bit as harrowing as on the Western Front, with the added discomforts of blistering heat and disease. As the scale of the disaster became sickeningly apparent, both Churchill's and Fisher's jobs were on the line.

The careers of these two colossi had been intertwined for years. Churchill admired Fisher's dynamism and innovation, while Fisher found Churchill 'Napoleonic in audacity, Cromwellian in thoroughness', and considered him 'a staunch friend'. Nevertheless, the cracks were beginning to show. After the war, and with his characteristic literary flair, Churchill compared Fisher to 'a great castle that has contended with time'. Hall, more prosaically, agreed that Fisher was 'a tired man. The strain under which he worked would have been terrific … He might still on occasion show the old flashes of brilliance, but, beneath the surface, all was far from being well … at any moment, we felt, the breaking point would come.'

At the same time, Hall was acutely aware that Churchill 'had the defects of his great qualities. He was essentially a "one man show". It was not in his nature to allow anybody else to be the executive authority.' He recalled an incident where they were debating an issue from diametrically opposed points of view: 'it was long after midnight and I was dreadfully tired, but nothing seemed to tire the First Lord. He continued to talk and I distinctly recall the odd feeling that although it would be wholly against my will, I should in a very short space of time be agreeing with everything he said.' Battered into submission, feeling his own sense of self diminishing rapidly, Hall admitted defeat: 'I can't argue with you; I've not had the training.'

*Admiral Jackie Fisher, 1914*

Not surprisingly, both Churchill and Fisher had their enemies. Working in harmony, they could deflect the critics and backbiters, but once serious differences over the Dardanelles campaign surfaced, they became vulnerable. In a fit of pique, Fisher offered his resignation. Churchill, knowing that this could spell doom for him as well, immediately wrote to Fisher on 15 May: 'if you go at this bad moment … thereby let loose upon me the spite and malice of those who are your enemies even more so than mine'. Fisher was unmoved; he reminded Churchill that he had been 'dead set against the Dardanelles operation from the beginning' and reaffirmed his decision to 'GO'.

At this critical juncture, Hall was asked to act as assassin by Sir Frederick Hamilton, Second Sea Lord. Senior figures within the Admiralty felt the time had come for both Fisher and Churchill to step down: the Dardanelles campaign had been Churchill's baby, which he'd pushed through despite serious concerns about its value. Now, as the hard-pressed British were having to commit more and more resources to landings at Gallipoli that had only become necessary because of the failure of the navy to do its bit, Churchill had nowhere to hide. Given that the ageing and autocratic Fisher had returned to the top job because Churchill had insisted on his appointment, it was inevitable that he would be ousted as well.

The problem was how to remove them without causing a backlash within the service and accusations that they were being sacked because of who they were rather than what they'd done. Hamilton needed to find someone of sufficient authority and unquestionable integrity to set the wheels in motion. Hall was an obvious choice. Nobody would accuse him, as Hamilton deftly put it, of acting through 'motives of self-interest'. As Hall was no longer on active service, commanding a ship, he had nothing to gain from a change of command. Hamilton approached Hall and asked him 'to take such steps as will make it impossible for Lord Fisher to ever return to the Admiralty. I consider him a real danger.'

Hall reluctantly accepted this 'most unpleasant job'. But to whom should he present the poisoned chalice? Showing the tactical awareness that served him so well in the corridors of power, he decided to put the case before Lord Reading, the Lord Chief Justice, as Asquith, the prime minister, who would ultimately be responsible for the decision, 'relied at this time on his advice'.

The two heavyweights met at Lord Herschell's flat. According to Hall, 'at 3.30 Lord Reading was shown in, and I at once put all my cards on the table. I told him I knew I was putting my entire future in his hands, but hoped to make him understand I was acting solely in the public interest … I said bluntly that in my opinion Lord Fisher was in no fit state to continue at his post.' Hall spoke for ten minutes: 'when I had finished he cross-examined me for nearly half an hour … question, indeed, followed question, some purely technical but others so fashioned as to make sure of my motive'.

Reading asked if it should be Fisher or Churchill who should go. Hall did not hesitate and replied, 'both … you can't keep a First Lord who will appear to have driven out of office a man like Lord Fisher. The navy would never forgive him.' Satisfied, Reading passed judgement: 'if you had answered my questions differently I would have broken you. But I am now satisfied that your view of what is required is correct and I will see the PM at once.'

In a matter of days both Fisher and Churchill were gone. Though Churchill refused to hold a grudge against Hall as the man who helped pull the trigger, Fisher was not so forgiving. In a letter dated 27 January 1917, he referred to Hall as 'that blinking rogue'. Later that year he accused Hall of being 'the champion liar of the British navy' and demanded he be replaced, because 'the NID should be thinking about our enemy's plans and what he is up to, and not hunting spies and interviewing journalists as its prime occupation'.

Though Fisher's accusations were motivated by personal animosity, there was some truth to them. Hall *was* in constant contact with the press. The print media played an aggressive role during the war, particularly the Northcliffe press, which was masterminded by the newspaper baron Lord Northcliffe and included *The Times* and the *Daily Mail*, Britain's best-selling tabloid. Sharp with their criticisms and ruthless in pursuing those who they felt were undermining the country's war effort, the barons of Fleet Street wielded their power mercilessly.

Hugh Cleland Hoy, who served in close proximity to Hall for most of the war, described the 'working agreement ... made between the Chief Censor and Admiral Hall early in the war in regard to the issue of news to the most important papers, who from the beginning included many Americans'. To deal with journalists, 'Hall evolved a system of entertaining them to tea once a week, when he would give a general resumé of the week's achievements. Admittedly, he did this with discretion, not to say bluff, at which he was a past master.'

At the same time, Hall was not above manipulating the press for his own ends. He realised he could exploit the fact that 'most foreign countries have the highest respect for our leading newspapers and believe what they read in them'. Given this, 'an intelligence officer can have fairly free scope', though he was careful not to compromise the reputation of the quality papers: 'whatever information you ask them to publish must be well within the bounds of reason'. He recognised, however, that 'with the penny press one need not be so careful. They are quite accustomed to eating their words and digesting the result.'

During the autumn of 1916, Hall hit on a way of taking the pressure off the exhausted British troops on the Somme. He decided to float the idea that an invasion of northern Belgium was being prepared, thereby drawing German forces away from the front line. The first step in this ruse was to send wireless messages to 'various stations' indicating that an invasion fleet would depart from Harwich, Dover and the Thames.

He then persuaded his closest ally in the press, Thomas Marlowe, editor of the *Daily Mail*, to insert 'a wily paragraph' in the paper that would support the invasion story. This was not straightforward: Marlowe couldn't simply print it in a regular UK edition, as it would be removed immediately by censors and never see the light of day. Instead, on 12 September, he had 24 copies of a special issue of the paper shipped to Holland for the benefit of German spies there. Most had the relevant paragraph blacked out to convince its German readers that it was a genuine edition that had already been scrutinised by the censors, leading them to assume that the few copies that still carried the offending article had somehow slipped through the net.

The paragraph in question was by a special correspondent and bore the headline 'EAST COAST READY. GREAT MILITARY PREPARATIONS'. It dropped heavy hints about 'very large forces concentrated near the East Coast'. In fact the preparations were on such a scale that they were unlikely to be for mere defence. The article worked a treat and the Germans did exactly what was expected of them: a mass of troops was moved to defend the Belgian coastline.

However, this sudden deployment sparked a major panic in Britain and caused the most serious invasion scare of the war. Hall remembered how 'our trenches along the East Coast were hurriedly manned; local commandants worked 24 hours a day; orders for the evacuation of residents in all towns and villages near the sea were on the point of being issued'. Perhaps wisely, he never perpetrated such a big deception again. Nonetheless, he was not deterred from making use of the press whenever it suited him.

The other activity that Lord Fisher had accused Hall of wasting time on was 'hunting spies'. Hall's investigative co-pilot on these missions was Basil Thomson, head of both Special Branch and the Criminal Investigations Department (CID). The origins of Britain's secret police lay in the 1880s, when the Special Irish Branch was formed to combat republican terrorists, while CID concentrated on the threat of anarchist subversion. Though MI5 was charged with uncovering spies operating in the UK, it had no powers of arrest; that honour belonged to Special Branch, and Thomson exploited it to the full.

Thomson was born in 1861. His father was the Archbishop of York, and he grew up in a restored fourteenth-century mansion. A sensitive youth, he suffered two breakdowns, the second forcing him to leave Oxford after only one term. Declaring a desire to travel, he was sent to an agricultural college in Iowa that turned out to be nothing but a farm set in the vast, unforgiving wilderness of the American Midwest, populated by wolves and rattlesnakes. Not surprisingly, he succumbed to nervous exhaustion.

For a man of his class and background, there was always the Colonial Office: Thomson was dispatched to the South Sea Islands in the Pacific. Despite enduring volcanic eruptions, cannibals, a war with a pirate tribe, and malaria, he enjoyed it immensely. In 1890, he became Commissioner of Native Islands and adviser to the King of Tonga. The first white man to learn Tongan, he was invited to join the King's family; as he put it, 'at the age of twenty-nine to be elder brother to a monarch of ninety-two is an unusual experience'. He was also elected prime minister, and administered Tonga's legal system, civic codes and tax affairs. Some years after leaving the island, with Germany casting a covetous eye on the region, he returned to negotiate its formal entry into the British Empire.

Back in the UK, he was a changed man; gone was the scared, overwrought schoolboy. For a while he looked after the interests of the crown prince of Siam and his brother; then, in 1896, after passing the bar exams, he was appointed deputy governor of Liverpool Prison. Thomson hoped he'd 'find criminals of our own race even more interesting than alien islanders on the threshold of civilisation'. In 1900 he transferred to Dartmoor Prison, where, in the wake of a riot and several escape attempts, he introduced a farm and a factory to give inmates something useful to do. His reputation as a troubleshooter followed him to Wormwood Scrubs, which had recently suffered a mutiny. Then, in June 1913, the position at Scotland Yard became vacant and he was asked to fill it.

Together, Hall and Thomson were a formidable team; during interrogations they adopted classic good cop/bad cop tactics. Hall's 'shrewd penetrating glance' and 'quick mode of speech' created 'an almost terrorising effect'. Thomson, on the other hand, was 'charming, quiet and sympathetic'.

One of their main tools for tracking and dealing with suspect individuals was set up in August 1915. The Port Control Section inspected the documents of everybody arriving in the UK by boat and liaised with the Military Permit Office, which had bases in London, Paris, Rome and New York and was responsible for issuing entry and exit visas. Information gathered by both agencies was sent directly to Hall and Thomson, who could then check the names against their own databases and act accordingly.

Aside from the titled German agents fleeing American justice, many of those stopped this way proved to be harmless amateurs who posed no real danger. This was not true of Alfred Hagn, a Norwegian traveller posing as a businessman. Even before he arrived, Hagn was on Hall's radar: 'the decoding of wireless messages revealed that a man of this name had left Holland for England ... the man's name already figured in our card index'. Hall wanted him arrested immediately he stepped off the boat; Thomson preferred to let him land and then have him followed. Hall reluctantly agreed, stressing that 'he must not be lost sight of for a moment'.

As Hagn made his way to London, he was shadowed by an undercover Special Branch operative who observed how protective he was 'of one suitcase in particular'. Hagn booked into the Savoy, and the Special Branch man set about befriending him. After a few days he concocted a way to get Hagn out of the hotel for long enough to search his room. In the suitcase, the agent discovered 'the materials of a fiendish plot', including phials, packages and documents. On Hagn's return to the Savoy, he was arrested.

When the contents of the suitcase were examined, it was found that one of the phials contained 'enough germs ... to kill thousands of people'; there were also 'enough explosives to reduce a large part of London to ruins'. Interrogated by Hall and Thomson, Hagn confessed that he intended to poison reservoirs and 'blow up vital centres in the London area'. Hagn escaped the hangman's noose: his death sentence was commuted to penal servitude and he was repatriated to Norway in 1919.

Of equal significance to his alliance with Thomson was Hall's relationship with Mansfield Smith-Cumming, head of MI6. The Secret Service Bureau was formed during 1909 to counter the perceived threat of German espionage, whipped up by the press and flamboyant, bombastic authors like William Le Queux, who penned lurid tales of spy rings and invasion plans. The bureau was split between MI5, whose remit was domestic counter-espionage, and SIS, later MI6, which was responsible for foreign affairs. As MI6 was under the nominal control of the Admiralty, a naval man was chosen to lead it.

Mansfield Smith-Cumming, dubbed 'C' by his colleagues, was born in 1859. He was educated at the Royal Naval College, Dartmouth, and at the age of 19 went to sea, spending six years patrolling the East Indies, where he saw action in operations against Malay pirates. He was decorated for his role in the Egyptian campaign 1882, but retired three years later due to ill health. In November 1898 he was put in charge of the construction of harbour defences at Southampton, and bought a country house not far away where he kept his collection of sailing vessels.

Cumming loved speed and enthusiastically embraced the new, revolutionary modes of transport that appeared with stunning regularity at the turn of the century. He joined the Motor Yacht Club, which held international races, and in 1906 was a founder member of the Royal Aero Club, established just three years after the Wright brothers made their historic flight, earning his pilot's licence in 1913. After the war he kept a biplane and a functioning tank on his lawn.

Like Malcolm Hay, he had a passion for fast cars; a friend remembered how he drove 'at breakneck speed about the streets of London to the terror of police and pedestrians alike'. He joined the RAC in 1902, and a year later drove a 50 h.p. vehicle in the Paris–Madrid race, during which eight people died. He gave a vivid account of the thrills and inherent danger of the early days of motor sport: 'the dust when passing other cars was awful. The mere physical pain of the stones hitting me in the face was considerable – my goggles were smashed ... the dust was so thick that one could not see the road at all.'

His love affair with the motor car ended in tragedy. In October 1914, while driving from Paris to Rouen with his 24-year-old son Alistair, the car skidded and smashed into a tree, crumpling on impact. Alistair was killed outright and Cumming's right leg became trapped in the wreckage. A grisly version of what happened next gained credence thanks to the well-known author Compton Mackenzie, who worked for MI6 during the war. He wrote that Cumming, penknife in hand, 'hacked away at his smashed leg until he had cut it off'. In reality, the leg was amputated in a French hospital soon afterwards. Cumming did little to deny the rumour that he had performed the surgery himself, it only added to his mystique. Once fitted with an artificial limb, he would, for his own amusement, stab it with a compass to alarm unsuspecting and unwanted visitors.

Though MI6 was technically under the control of the Foreign Office, it found itself under constant pressure from the War Office and Military Intelligence, who were running their own agents in parallel to Cumming. Matters were further complicated by the fact that FO officials in target countries often resented the presence of spies, and the fledging MI5 under Vernon Kell was also trying to get in on the act.

These internal power struggles frequently led to confusion and messy compromises that threatened the very existence of the networks that had been painstakingly set up by MI6. As a result, intelligence-gathering in Switzerland was almost entirely abandoned; in Holland and Belgium, two critical territories, it led to the collapse of espionage operations; while in

Russia, personality clashes were made worse by the fact that the Russians' own agencies were more interested in spying on each other than on the enemy.

Cumming's one powerful friend in all this was Hall. At various points throughout the war, the WO made determined efforts to bring MI6 under their control. Each time, Hall stood by Cumming and insisted on his independence. Cooperation between the two even extended to codebreaking. Cumming had a very small unit, mainly creating codes for his own agents to use. Room 40 staff often dropped by to swap ideas and give advice.

In return, Cumming was expected to serve Hall and support his intelligence operations. Over the course of the war, he ran agents for Hall in Spain, Greece, Romania and South America. Hall also insisted that Cumming's agents in Germany, who were few and far between, report back on naval matters such as construction and repair work going on in the dockyards there. In addition, he had Cumming employ watchers on the Danish coastline to spy on the movements of the German fleet and monitor neutral ships that might be carrying contraband in defiance of the blockade; the same arrangements applied to Norway and Sweden.

Though Cumming had little choice but to bow to Hall's demands, he could be fiercely protective of his favourite agents. One of them was the prolific young writer Compton Mackenzie, best known today for his Scottish novels, *The Monarch of the Glen* (1941) and *Whisky Galore* (1947). Mackenzie had been working for MI6 in Greece and the Aegean, but, increasingly frustrated at the incompetence and lethargy he encountered, he wrote a critical report about the situation and forwarded it directly to Hall rather than sending it to Cumming, after a colleague advised him that 'Blinker ... holds senior rank to him in the service'. When Cumming heard about it, he was livid with Mackenzie: 'I regard your behaviour in sending a report ... over my head to my superior officer as a gross breach of discipline.'

Back in London, Mackenzie was hauled before Cumming. He feared the worst, but Cumming's bark turned out to be worse than his bite. After ignoring Mackenzie for a few moments, he 'took off his glasses and stared hard at me for a long minute without speaking', then demanded he account for himself. Mackenzie spun a yarn, 'in such a way as to win his attention', and they embarked on a two-hour conversation that ended in an invitation to dinner. Cumming paid for the taxi that Mackenzie had left waiting outside throughout the meeting, having anticipated that it would last only a few minutes, and they went to eat. During the meal, he introduced Mackenzie to his wife as 'the man who had given him more trouble than anybody in the service'. He confessed that he had intended to make life extremely unpleasant for Compton, but instead found him to be 'a man after my own heart' and suggested he visit every day he was in London.

In autumn 1917, Mackenzie returned to the Mediterranean, spending some time in Italy, and passed through Paris, where he managed to ruffle the feathers of an FO official, resulting in a mild and wryly amused telling-off from Cumming. Despite the trouble Mackenzie caused him, Cumming wanted him to become his number two. Mackenzie declined the offer and headed to the island of Capri, where he started on a new book.

*The writer Compton Mackenzie in uniform*

Rumours that Mackenzie was writing about his espionage work soon reached Hall. 'Boiling' mad, he summoned the young man to the Admiralty. Mackenzie recalled the meeting in some detail: 'I had heard a great deal about Blinker Hall … what I had not heard was that facially he was so like my own chief. His nose was beakier; his chin had a more pronounced cutwater. Nevertheless, when I looked at the two men I could have fancied that each was a caricature of the other.'

Hall got right to the point and asked him about the book, reminding him that he'd be liable for a court martial, and to forget about getting it published in America as a way round censorship. Mackenzie denied he had any such intention and assured Hall that he would only pen an account after the war was over. Hall was not convinced. 'I'm talking about while the war lasts. I hear you've already written at least half a dozen chapters.' He paused to fix Mackenzie with a 'horn-rimmed horny eye. I felt like a nut about to be cracked by a toucan.'

Mackenzie withstood Hall's scrutiny, insisting that it would take years for him to process his experiences and turn them into the written word. The interview ended, and though Mackenzie doubted whether Hall believed him, Mackenzie felt that he 'thought he had frightened me out of persevering with this mythical work'. As it was, Mackenzie returned to Capri and his writing and played no further part in the war, his absence from duty justified on health grounds and fully supported by Cumming, who never lost his affection and respect for the troublesome Mackenzie.

Considering that Hall was the official guardian of one of Britain's most closely kept secrets, the existence of Room 40, it could be argued that by making himself such a ubiquitous presence in the corridors of power and beyond, he increased the chances that it would be discovered. In fact, the reverse is true. Hall was hiding in plain sight. The more attention he attracted, the more he deflected it away from the codebreakers. He put himself at the forefront of the intelligence effort not only because it was in his nature to do so, but also because it made him a more effective gatekeeper.

Hall was determined to keep Room 40 under wraps, admitting that 'there were several occasions when I made myself highly unpopular by refusing to divulge information to those who, perhaps not unreasonably, considered that they ought to have been given it'. After all, if the Germans got the merest hint of Room 40's existence, they would instantly change all their code books, tighten the security of their communications, and monitor their use of wireless more carefully. All that Room 40 had gained would be lost.

Hall did not even trust other government departments with knowledge of Room 40's existence. Outside of the Cabinet and a few high-ranking civil servants, it remained a secret, one Hall would go to considerable lengths to protect. A prime example of his intransigence was recorded by Basil Thomson: a naval man who 'deserted in Spain and went to a German spy to give him all he knew about British naval movements' was arrested after 'the message sent … was intercepted by the Admiralty'. To keep Room 40's role in the affair secret, Hall had the man charged with desertion, rather than colluding with the enemy. However, 'through a blunder', the culprit was allowed to go home to Barrow and get a job with Vickers, the armaments manufacturer.

Incensed, Hall had him arrested again and taken to London to be interned. At this point, the War Office got involved. They wanted the man set free 'unless they were put into full possession of all the facts of the case'. Hall refused as it would mean revealing Room 40's involvement.

Next, Hall approached the Home Office. Unfortunately, they asked for the same degree of corroboration, unsettled by the fact that the man's fellow workers at the Vickers factory were threatening to go on strike if he wasn't released. With nowhere else to turn, Hall called in Thomson, who interrogated the suspect and came to the conclusion that he was probably insane. Though it went against all his instincts, Hall let the matter drop. Maintaining the security of Room 40 was vastly more important than securing the conviction of one deluded sailor.

## Chapter 10

# MEDITERRANEAN THEATRE

After Italy entered the war on the Allied side in the spring of 1915, Spain, Portugal and Greece were left clinging precariously to their neutral status. Each country faced similar problems: industrialisation had given rise to a militant left backed by powerful trade unions, opposed by entrenched conservative interests supported by the army and the Church, while weak liberal democratic governments came and went with dizzying rapidity. A mood of crisis was all-pervasive. Italy was in the same boat but took the plunge anyway.

Had the war ended as quickly as expected, the other Mediterranean countries might have been able to avoid too much collateral damage. As it was, the pressure increased for a commitment either way, resulting in deeply damaging splits that further destabilised them, as the Great Powers, desperate for any advantage that might break the deadlock, applied the thumbscrews.

Portugal succumbed first. In January 1915, a pro-German military coup unseated the government, followed a few months later by a popular rebellion against the new regime. When the tiny German force in East Africa, which had been evading all British attempts to run them down, crossed into the Portuguese colony of Mozambique, a decision became unavoidable and Portugal declared war on Germany in March 1916.

Hall's main focus was on Spain, which had quickly became a political and diplomatic battleground and a front line in the espionage war. The Germans were favoured by the right-wing Carlists, while the republican radicals were pro-Allies, leaving the liberal government stranded somewhere in between. Aside from the country's geographical location – its coastline offered access to the Mediterranean and the Atlantic – Spanish exports of iron ore, coal and fruit, particularly oranges, were of vital importance to the combatants.

Determined to have his own way, Hall took control of all covert activity in Spain, running agents and coordinating operations. As a result, much of the work carried out by Room 40's new diplomatic section was focused on the German embassy in Madrid, which not only managed a large network of spies across the country, taking in all the major cities and the island territories, ably supported by the 70,000 German citizens living in Spain at the time, but also acted as a clearing house for communications with agents in other countries.

At the heart of this espionage set-up was the naval attaché, Baron Hans von Krohn, an unpleasant zealot who advocated using cholera bacilli to poison the main rivers on the Spanish–Portuguese border; this idea proved too much even for Berlin to contemplate and was rejected outright.

The Berlin–Madrid axis provided Room 40 with a treasure trove of information. When, during 1915, the Germans wrongly suspected that their messages to Madrid were being picked up by the French due to the work of either a spy or a traitor, they attempted to increase the security of their codes by introducing three new cipher keys. Unfortunately for them, the message containing these new keys was sent by wireless and fell straight into the lap of the grateful British codebreakers.

With Room 40's intelligence to guide him, Blinker Hall looked around for somebody who could be sent to Spain to do his bidding. In an inspired move, he chose the well-known author A. E. W. Mason. This extraordinary man resembled the heroes of his many action-adventure novels and he performed his duties with relish, vigour and a devil-may-care attitude. As he saw it, the secret service was 'a service with its traditions to create. Indeed, it had everything to create, its rules, its methods, its whole philosophy. And it had to do this quickly during a war.'

Born in 1865, Mason went to Oxford, where he became involved in theatrical and debating societies. After graduation, he embarked on an acting career and did a couple of years in rep before deciding to become a writer, alternating between stage plays and novels. He wrote a series of period romances which sold well, then produced his masterpiece, *The Four Feathers* (1903), a stirring imperial adventure. His reputation as a literary star was sealed by his West End hit *The Witness for the Defence* (1911), a courtroom thriller.

Meanwhile, he lived life to the full. He travelled widely, visiting North Africa, South America, India and Afghanistan. He was immensely sporty, taking in golf, fishing, yachting, cricket and mountaineering. Though never married, he had a string of lovers and a long-running affair with an American actress. A natural bon viveur, he was, as one of his friends noted, 'the ideal club man ... as good a listener as a talker', with 'an appreciation of food, drink and tobacco'.

Pushing 50 in 1914, he was initially part of a committee of literary heavyweights, including Arthur Conan Doyle, Thomas Hardy and H. G. Wells, set up to produce propaganda material, but Mason had itchy feet. When he got the call from Hall in early 1915, he leapt at the chance to be part of the action.

In assessing Mason's espionage career, it is difficult to untangle fact from fiction; in his writings they are completely interchangeable. A compulsive storyteller, he couldn't resist the urge to get his experiences into print as quickly as possible, including transparent translations of many actual events and the people involved, with the main characters functioning as his alter egos. He did leave notes for an autobiography, which contained some of his field reports, while Hall, when he was putting together his own unpublished memoirs, asked Mason to supply details of 'three or four incidents which I would particularly like to bring out'. These two sources offer solid confirmation of some of the fictionalised episodes that appear in Mason's books.

In *The Summons* (1920), Mason is represented by the playwright Martin Hillyard. When war breaks out, Hillyard is called in to see Commodore Graham (Hall), who had 'a whole array of cipherers and decipherers ... in different rookeries in London. Commodore Graham's activities embraced the high and the narrow seas, great capitals and little tucked away towns and desolate stretches of coast where the trade winds blew.' Graham decides to send Hillyard on a cruise: 'in the end his finger rested on the name of the steam yacht *Dragonfly*, owned by Sir Charles Henderson'. In reality, the yacht in question was the *St George*, owned by Lord Abinger. Mason was to assume the role of an eccentric millionaire, a 'mad Englishman'. He cruised round Spain, Gibraltar, Morocco and the Balearics, hosting a non-stop house party. A permanent guest was W. E. Dixon, Professor of Pharmacology at Cambridge, who appears in the novel as Paul Bendish, an expert in codes and secret writing.

A U-boat encounter in 1916, 'The Glass Tubes Affair', was mentioned in Mason's operational notebooks and referred to by Hall as 'the incident of U-35 dropping the cases in Cartagena harbour'. These cases were packed with glass tubes containing hydrofluoric acid and were heading for France and England. The tubes were then going to be smuggled into various munitions factories and carefully hidden near explosive material. The acid, acting like a time fuse, would take four and a half hours to eat through the glass before setting off a chain reaction that would set the factory ablaze.

An account of how Mason was able to discover the details of U-35's mission is included in *The Summons* and includes the tale, probably true, of the secret communication method used by a German agent in Barcelona who had his letters addressed to non-existent English people and bribed the postman to deliver them to him. Letters containing secret material were written in invisible ink. Mason's fictional doppelgänger, Hillyard, gets hold of one and uses iodine to bring out the hidden message: 'a submarine will sink letters ... and a parcel of tubes between the twenty-seventh and thirtieth of July, within Spanish territorial waters off the Cabo de Carbon. A green light will be shown in three short flashes from the sea and it should be answered from the shore by a red and a white and two reds.' Hillyard waits on a deserted beach on the relevant night, makes the necessary signals and picks up the consignment himself.

Mason also foiled German attempts to encourage rebellion in French Morocco. Operating from Spanish Morocco, the German agent on the ground was trying to arm and finance rebel attacks on the French colony. In early 1916, Mason wrote in an official report that he 'crossed to Tangiers on

13 February in order to discover ... what the Germans were really achieving in the Riff country and by what means we could best attack their influence there'. He managed to identify the local banker who was handling the flow of funds to the rebels, and was able to cut off their finance at source. He then turned his attention to the supply of weapons. In his short story 'Peiffer' (1917), he mentions foiling German plans to smuggle ammunition and 50,000 rifles into Morocco, while in the novel *The Winding Stair* (1923) he describes how the intelligence he produced led to the destruction of a major gun-running operation.

In this case, Room 40 decodes support Mason's fiction. A message from 7 October 1916 revealed that the rebels were ready for action but chronically low on ammo; three weeks later, another intercepted message reassured them that seven men, four machine guns, 1,000 rifles and 50,000 francs were on their way. Within days, a million rifle cartridges were seized by local police at Madrid railway station.

According to Cleland Hoy, Room 40 'decoded many cipher messages that were passing between Morocco and Berlin', including requests for more armaments: the weapons successfully made their way to the U-boat which would deliver them to Morocco. Having informed the French naval attaché at the Admiralty, Room 40 waited for further developments: 'at last over the ether came the all-important message with the news that the U-boat had set out' loaded with its deadly cargo. The submarine was followed by French planes, bombed and sunk off Larache, consigning the rifles and ammunition to the bottom of the sea.

In January 1917, fearing that his cover as a playboy yachtsman was close to being blown, Mason returned to England. As he explained in *The Summons*, 'the purpose of the yacht was long since known to the Germans. The danger of torpedo was ever present ... any means would be taken to force him to speak before he was shot ... he carried hidden in a matchbox a little phial, which never left him.' Mason did possess such a phial, probably cyanide, which he kept in a secret compartment in his desk for years after the war and often showed to his friends.

*

At the end of 1917, having been home for a number of months, Mason, whom Hall regarded as a prize asset, was in Mexico, a hornets' nest of German intrigue. His replacement was a much less flamboyant character. Charles Thoroton, a Royal Marines colonel, was a dedicated professional who helped neutralise a major plot to transport wolfram from Spain to Germany. Mined in the Basque region, wolfram, also known as tungsten, was an iron ore used to create an alloy that made explosive projectiles – shells, shrapnel, grenades – considerably more effective.

During September 1917, Thoroton's man in Bilbao reported that German agents were planning to smuggle large quantities of the precious metal out of the country. On 2 October, Room 40 decoded a message from Berlin to Baron von Krohn, the naval attaché and spymaster in Madrid, about the possibility of shipping wolfram by submarine: 'the execution of this plan is perhaps possible in November in the neighbourhood of the Canary Islands'. Krohn replied on the 16th agreeing that it could be done. Later that month Thoroton's agent located the warehouse near Bilbao where the wolfram would be stored before departure, and identified the ship, the *San Jose*, that would carry it.

However, a few weeks later, the *San Jose* sailed without the ore. A different ship, the *Erri Berro*, had been selected for the mission. This was confirmed by Room 40, as was the procedure it would follow: 'the sailing vessel will receive sealed instructions concerning recognition signals. When sighting a U-boat at the rendezvous she will lower and unfurl her sails and hoist a Spanish flag as well as blue and yellow pennants under one another. At night she will show a blue and yellow light.'

At first Hall was adamant that the ore should not leave Spain, but he swiftly changed his mind after Room 40 decoded another message from Berlin to Krohn: 'U-cruisers 156 and 157 can be at the Canaries on November 24th. Each can take about 40 tons of wolfram ore … please report details and meeting place can be settled.' Two further messages gave the rendezvous point, code name U Platz 30.

Hall realised that this was a chance to get not only the wolfram, always useful and always in short supply, but the U-boats as well. A naval operation was put in motion to destroy them, approved by Jellicoe and the French naval commander-in-chief, who had overall responsibility for the Mediterranean theatre. Four E-class British submarines were assigned to the U-boats, while HMS *Duke of Clarence* was to intercept the ore shipment.

The German U-boats, U-156 and U-157, began their journey to the Canaries on Christmas Day. Their delayed departure was due to problems with the *Erri Berro*, which, after changing crew and having repair work done, finally set forth on New Year's Eve at 4 p.m. Two hours later the *Duke of Clarence* was on its way to intercept it. Half an hour into the New Year, it captured the *Erri Berro* with the wolfram on board. However, as tow lines were being attached to the offending vessel, *Clarence* bumped into it, inflicting irreparable damage; as evening fell, the British were left no choice but to scuttle it. The Spanish sailors were taken prisoner and sent to a detention centre in South Kensington.

This still left the matter of the U-boats. E-35 and E-48 arrived at the meeting point, close to the Canaries, on 30 January. U-156 was waiting, having spent the intervening period bombarding Portuguese coastal towns. The British submarines fired three torpedoes. Two missed; one hit but failed to explode. The other U-boat, U-157, never showed at all. Both it and its sister ship made it safely back to Germany.

Hall, who mostly kept his volcanic temper in check, was so annoyed that when he returned home to find his wife entertaining some ladies to tea, he kicked over a table laden with sandwiches and stomped out without a word. Some compensation for the escape of the U-boats came in the months that followed. Once the Spanish crew of *Erri Berro* were repatriated, the story appeared in the Spanish press and the resulting scandal ended with Krohn being expelled from Spain.

Before Krohn was unceremoniously run out of town, he'd been a firm advocate and facilitator of biological warfare. By 1917, Room 40 had picked up a stream of messages suggesting that Spain was a transit point for anthrax and glanders bacilli. Carried by submarine from Austria's Adriatic bases, the diseases arrived at Cartagena harbour, where they were unloaded and then shipped by liner to South America to be injected into mules and cattle destined for the Allies. Mason provided the first solid proof of this nefarious business: he stopped a consignment of anthrax that had been hidden in shaving brushes.

More was to come. During February 1918, Thoroton, acting on Room 40 decodes, tipped off the local chief of police at Cartagena and U-35 was caught with 12 cases of anthrax and glanders concealed in lumps of sugar. The fact that this horrific cargo was travelling under orders from the German embassy presented Hall with an opportunity to embarrass the German ambassador, who was already smarting from the dismissal of Krohn.

To perform this delicate mission, Hall employed his personal secretary, Lord Herschell. As lord-in-waiting to both Edward VII and George V, Herschell was extremely well connected and was friends with the Spanish monarch, King Alfonso XIII. He presented the damning evidence to the King, and, shocked by this blatant abuse of diplomatic privilege, Alfonso politely told the German ambassador that he was no longer welcome in Spain.

Support for Germany was wearing thin in Spain. Between April 1917 and April 1918, in a desperate effort to disrupt trade with the Allies, U-boats sank 40 Spanish merchant ships and killed 100 sailors. For the same reason, German agents promoted labour unrest, strikes and industrial sabotage. By 1918, they were conspiring with anarchist groups as well as recruiting informers and crooked cops, including the head of Barcelona's political police. These murky dealings culminated in the assassination of José Barret, a major player in the Catalan metallurgy industry, whose factories supplied the French with shells. The investigation into this brutal murder implicated government officials and local dignitaries. Shocked into action, politicians passed an espionage bill to crack down on German spies.

As the war drew to a close, Hall's agents were paid the highest possible compliment by the Spanish government: it begged him not to withdraw them because, according to Edward Bell, Hall's confidant at the US embassy in London, they were 'a far more reliable source of information ... than their own police and civil authorities'.

*

King Constantine, the Greek monarch, was pro-Germany. The prime minister, Eleftherios Venizelos, who'd been elected in June 1915, was pro-Allies. The potential for conflict between them increased when Bulgaria entered the war in October 1915 and helped the Germans and Austrians crush Serbian resistance. The Allies, bogged down at Gallipoli and fearing that the strategic balance in the Balkans was going against them, asked Venizelos if the surviving remnants of the Serbian army could seek refuge in the Greek territory of Salonika.

Venizelos agreed to their request, but when news of the deal came out, his government collapsed and he was forced to resign. A new administration, supported by the King and preaching benevolent neutrality, rejected the Salonika plan. Ignoring their wishes, and with Venizelos still agitating on their behalf, the Allies went ahead anyway and began landing a force of 75,000 British and 75,000 French troops at Salonika.

Given the uncertainty surrounding Greek policy, it was imperative that MI1(b) tackle their diplomatic codes; Malcolm Hay remembered that 'in 1916 information about what was going on in Greece was badly wanted'. MI1(b) had access to 'very long messages which were passing in great numbers between King Constantine ... and Berlin', but there was a major stumbling block. Hay had no means of knowing in what language the messages were written.

His right-hand man, John Fraser, was given the job of solving the riddle. A lecturer in Latin at Aberdeen University and later Professor of Celtic at Oxford, Fraser mastered 21 languages by the end of the war and was instrumental in cracking the codes of 11 different countries.

After several weeks of intense study Fraser concluded that the text must not be in Greek, but in French. This was the breakthrough they were looking for. Fraser immediately telegrammed Hay, who was out of the office for a few days, the simple message 'Pillars of Hercules have fallen'. After that, progress was swift. Fraser applied his discovery to a number of different code books, some in Greek, some in French, and, with Room 40's cooperation, set about reconstructing them.

By the end of the year, MI1(b) had the inside track on the deteriorating political situation in Greece. Venizelos continued to press for a clear commitment to the Allies. The King resisted as long as he could before being forced to abdicate in June 1917. Venizelos began to mobilise the army so it could lend support to the Allies at Salonika.

At the time, there were around 680,000 troops, British, Italian, French and Russian, rotting away on what had been dubbed the Macedonian Front. Two dismal and short-lived attacks were launched against the Bulgarians in the summer of 1916 and the spring of 1917; otherwise a dispiriting stalemate was the order of the day, only broken in the last few months of 1918 when the Allies went on the offensive and the Bulgarian army broke and ran, leaving its government no choice but to seek peace.

Malcolm Hay's conviction that MI1(b) should play a much greater role than it had under his predecessor led inexorably to an expansion of his team and larger premises to accommodate them. While he lived alone at 20 Gloucester Place, Hay's staff moved to a large building in Cork Street in the heart of the West End. The core codebreaking team grew from three to eleven: by the end of the war there were 20, plus 60 clerical and secretarial staff, almost all women. They had their work cut out, as Hay recalled: 'thousands of telegrams filled up the cupboards at Cork Street. Although my staff had increased it was impossible to read everything.'

Like Blinker Hall, Hay recruited mostly from academia: there was a medieval historian, a classicist, a lecturer in palaeography, an Egyptologist, an Arabist, an expert on Celtic languages, a philologist, the curator of the Ashmolean Musuem in Oxford, a mountaineer, and an ex-consul formerly based in Tokyo.

The security and privacy of MI1(b)'s operations was jealously guarded. To distract and detain unwanted visitors from neutral countries, a dummy room was maintained at the War Office. The entrance to Cork Street itself was guarded by 'a trusty warrant officer of prizefighter physique'; if anyone got past him, Hay had a colleague available to act as a decoy by giving 'the impression of a typical British idiot'. A letter to the police, regarding 'the nuisance caused by the large number of itinerant musicians' who congregated on the streets outside, politely demanded that the constabulary prohibit 'street noise in the immediate vicinity'.

The codebreakers at Cork Street focused mainly on diplomatic traffic. Unlike Room 40, which had been gifted copies of the relevant German code books early in the war, the decipherers at MI1(b) had to start from scratch. Hay noted that 'before decoding the messages, we had to reconstruct the code books'; many of them were non-alphabetical, adding to their complexity, while 'some embassies used to encipher their codes'. However, as he proudly stated, 'all these difficulties were overcome, Cork Street was never defeated'. By 1918, his team had broken the diplomatic codes used by the USA, Argentina, Brazil, Uruguay, Chile, Spain, Portugal, Switzerland, Norway, Sweden, Italy, the Vatican, Holland, Greece, Romania and Japan.

One of MI1(b)'s most significant achievements was cracking the 'Für GOD' system used by the German General Staff to encode messages sent three times a week to their secret agents in North Africa and the Middle East via the all-powerful Nauen transmitter. As the messages carried no signature, no address and no call sign, each normally a standard point of entry for the codebreakers, they had already defied Room 40's experts. The system was eventually solved by Captain Brooke-Hunt, one of Hay's key staff, who discovered that it contained 22 mixed alphabets and 30 random cipher keys of 11–18 letters, generating a dizzying array of variables and alternative letter combinations, making it extremely difficult to discover any repetitions or recurring patterns. The intelligence gained from this breakthrough was shared with Room 40, and Hall used it to stop German gun-running to rebels in French Morocco.

This was one of the few examples of the burgeoning cooperation between MI1(b) and Room 40 that we have on record. Unfortunately very little else remains to illuminate their relationship. Hay and Blinker Hall had a private telephone line installed and spoke at least once a day. Sadly, neither left any record of their conversations. A few documents in the archives feature requests by Room 40 for material from MI1(b) about

South American codes, the exchange of notes and code books relating to Spain, and some American material.

One such communiqué from Room 40 to MI1(b) concerned Oliver Strachey, who was by then overseeing its Middle Eastern work: 'touching on the Turkish question, I wonder if Strachey would like to see some intercepts of a year ago when Constantinople and Berlin were only joined by WT. As there is now an expert in this language something might in time be made of them which would throw light on subsequent events.'

This kind of exchange of material and ideas between Room 40 and MI1(b) reflected the informal camaraderie that developed over the course of the war, driven by a healthy sense of competition. It was in stark contrast to other sections of the intelligence community, which were beset by suspicion and barely concealed hostility. How much the diplomatic work carried out by Malcolm Hay's team dovetailed with Hall's efforts to manipulate the foreign policy of neutral countries and counter German efforts to do the same, we will never know. That said, the very fact that Hall and Hay brushed aside the mutual loathing that characterised the relationship between the Admiralty and the War Office demonstrated that their commitment to defeating the enemy took precedence over departmental politics.

# Chapter 11

# THE IRISH WAR

January 1916 was deceptively warm and dry in London, as if spring were coming early to assuage the bloody wounds of war. The British had just finished evacuating troops from the disaster of Gallipoli, after suffering close to 115,000 killed or wounded. On the Western Front, where the war was now entrenched into its second winter, the Germans and the French were about to begin the Battle of Verdun, the longest single battle of the war and one that would claim more than 700,000 combatants by the time it was finished – with no victor – in December 1916.

Blinker Hall was under no illusion that the war was going to get any easier, despite US President Woodrow Wilson dispatching his unofficial Secretary of State Colonel Edward House to Europe in early January to work out an Anglo-American strategy for peace. Though House spoke with French and German leaders about ending the war, he spent most of his time in London telling the British that the Germans would relaunch unrestricted submarine warfare, suspended in November 1915, and that this would bring the Americans into the conflict. But when?

And then there was the Irish question. In 1912, the Third Home Rule Bill for Ireland – violently opposed as 'Rome Rule' by the Ulster Unionists – had riled supporters and opponents on both sides of the Irish Sea. The bill's eventual success – it would receive royal assent in 1914 – led to sectarian violence and the simmering threat of civil war. But the Irishman who was most on the mind of Blinker Hall at the beginning of 1916 was in Germany, with a bold plan to open another front on which to further bleed the depleted British military. And he was a revolutionary from within the British establishment, which made him that much more dangerous.

Roger Casement, an elegant 50-year-old bachelor, had come to the United States in the summer of 1914 as a knighted servant of the British Crown who had won his title for his work in the British consular service in the Congo, and then in Brazil. Casement had done more than bid for the interests of the state in his 18 years in the Foreign Office. He had been a bold emissary of justice, investigating – and publicising – with zeal the crimes committed against rubber plantation workers by their corporate and government exploiters, especially the genocidal rule of Belgium's King Leopold II in the Congo, and the slavery and torture inflicted by the Peruvian Amazon Company on the Putomayo Indians in South America.

Casement had been born into a life of seeming Anglo-Irish comfort in Sandycove, just outside Dublin, in 1864. His father, the son of a bankrupt Belfast shipping merchant, was a captain in the 3rd Dragoon Guards of the British army, and a Protestant. His mother, a Dubliner, had her son secretly baptised as a Catholic when he was three years old. By the time he was 13, Casement was an orphan, and was taken in by an uncle, who raised him as a Protestant and sent him to boarding school.

The tall, handsome Casement cut a striking figure when he ventured into the Congo, as caught by the eye of Jessie Conrad, the wife of his friend, novelist Joseph Conrad: 'He was a very handsome man with a thick, dark beard and piercing, restless eyes. His personality impressed me greatly. It was about the time when he was interested in bringing to light certain atrocities which were taking place in the Belgian Congo.'

And indeed he did. In 1911, Casement named the perpetrators, many of whom were charged and convicted, and detailed their crimes in his report to Britain's Parliament on human rights abuses in the rubber industry, an account framed by the crimes he had seen in the Congo and reported on in 1904. His work won him a knighthood, and international celebrity as a humanitarian crusader. Yet despite his success in the Foreign Office, he was disillusioned by the sins of empire, and especially those he now saw committed by the British in his home country. He had become increasingly politicised by the vigorous and violent opposition that Protestant Ulster Unionists successfully demonstrated against Irish Home Rule in 1912, a form of Irish self-government still overseen by Britain, which the UK proposed as a solution to the Irish Question. Casement realised that he was on the other side – the side of total Irish liberation.

Delving into Irish history and studying the Irish language, he left the Foreign Office in 1913 and helped to found the Irish Volunteers, a republican organisation whose mission was to help usher in self-rule in Ireland. That same year, he visited Connemara, in the west of Ireland, whose poverty shocked him even more than the misery he had witnessed in his consular work in Africa and South America.

The following year, with both sides having formed paramilitary organisations, Casement was in charge of a committee to buy arms for the Irish Volunteers. With a budget of £1,500 to buy 1,500 rifles – in contrast to the funds available to the Protestant Ulster Volunteer Force, which purchased 35,000 guns and three million rounds of ammunition – Casement shipped arms to the Irish Volunteers on the yacht of novelist and Republican sympathiser Erskine Childers, who would serve with distinction in the Royal Navy during the war.

In the summer of 1914, Casement travelled to the United States at the invitation of John Devoy, the 72-year-old patriarch of Irish nationalism in the US. Devoy had been exiled there in 1871 as one of the 'Cuba Five', Irish revolutionaries who were released from British prison on the condition they did not return to England until their original sentences had expired. He had been sentenced to 15 years for treason in trying to organise an uprising of Fenian soldiers he had met while serving in the uniform of the British army. It was this idea that fuelled Casement's war against the English.

In a display of Irish republican enthusiasm that annoyed the British, Devoy had been welcomed to the United States at the very seat of national government: the House of Representatives. Irish nationalism was a strong force in American political life, buoyed by the promise of Home Rule that had first been mooted by the British more than half a century before the 1914 war began in Europe. After such a warm embrace in the new world, Devoy had no interest in returning to the old, save for his politics. He became a journalist for the *New York Herald*, and eventually, leader of the Clan na Gael, which was the American branch of the Irish Republican Brotherhood. Under Devoy's direction, Clan na Gael had risen to become the most powerful Irish republican organisation in the USA, and it gave Casement a mighty platform from which to advance his cause of Irish nationalism with money and men. And then the war gave him an even bigger one.

In September 1914, Casement attracted the attention of his government, in a way that would ultimately become fatal, when he wrote an open letter to the Irish people from New York. In it he urged all Irishmen to refuse to fight against Germany, and declared himself a founding member of the Irish Volunteers. The British Foreign Office suspended his pension, and MI5 opened a file on him.

For the British, hoping to convince the United States – by whatever means necessary – to enter the war on the side of the Allies, the Irish Question was especially sensitive. The Great Famine had ravaged Ireland in the mid nineteenth century, killing a million people and speeding another million to distant shores, many of them staying in the place where they'd

*Sir Roger Casement and John Devoy*

arrived in the United States: New York City. The large, and largely urban Irish population wielded considerable influence in President Woodrow Wilson's Democratic Party, where they favoured peace because it was good for the American economy – and the Irish working man. The British knew that some of the Irish in America saw the war as a chance for Ireland to free itself from Britain once and for all. It was a dangerous game, and Blinker Hall was keeping a close watch on Irish republican traffic, including Roger Casement in New York. And Hall had some powerful local help.

While in New York, Casement was lodged at the home of John Quinn, a second-generation Irish-American lawyer who, in addition to being influential in the art world as a collector and patron, was also deeply involved in the world of Irish politics. The dapper 44-year-old Quinn was an energetic supporter of Irish Home Rule, and would help write an Irish Home Rule Convention in 1917. In 1914, however, he was also working on behalf of the Allied cause, feeding intelligence to the British consulate at 44 Whitehall Street. Back in London, Blinker Hall learned everything that Casement was plotting in America.

On 10 August, the swaggering, impassioned Casement met with Germany's opportunistic military attaché Franz von Papen at the German Club in New York, along with the German ambassador Count von Bernstorff, Quinn, Devoy and others. Casement's war plan for Ireland so inspired von Papen that he sent a memo to Germany introducing Casement as the 'leader of all the Irish in America'. Two weeks later, Casement wrote to Kaiser Wilhelm himself, outlining the rich possibilities that the war presented to Irish freedom from English rule: 'Thousands of Irishmen are prepared to do their part to aid the German cause for they recognise that it is their own.' In his letter, Casement argued that of the 150,000 British soldiers taken prisoner by the Germans, 35–40,000 were Irish, and they should be separated and organised into a separate Irish brigade, ready to land in Ireland and fight the English.

The British naval blockade of the Atlantic made that kind of armada unlikely, but the Germans realised that even the possibility of an invasion of Ireland would stretch British resources. So too did the British, who were keenly monitoring von Bernstorff's communiqués with Berlin. Though Blinker Hall had his eye on Casement in New York in order to trail the dangerous renegade diplomat, Casement – having honed his survival skills in the brutal African and South American imperial jungles – managed to leave the teeming city undetected and sail for Germany via Norway on 15 October 1914 under the false identity of James E. Landy.

In Kristiania (Oslo), Casement had a narrow escape when his manservant – and possibly lover – Eivind Adler Christensen, a Norwegian sailor whom he had met in New York, was corralled by the British and offered money to hand his master over. Christensen would later recall that the British actually wanted him to kill Casement, but once again the Irish war missionary made his escape, landing in Berlin on 31 October. Sir Roger Casement, Knight of the British Empire, seeker of Irish freedom, had now irrevocably gone over to the side of the enemy.

Things began well for Casement's Irish Brigade plan, after enthusiastic meetings with Under Secretary for Foreign Affairs Arthur Zimmermann and Count George von Wedel, head of the German Foreign Office's English section. Von Wedel sent a memo to Chancellor Theobald von Bethmann-Hollweg recommending that Irish Catholic prisoners should be transferred to a special camp to undergo training by the unlikely team of Casement and Irish priests. On 20 November, Casement received a rousing public endorsement when the German government announced in the *Norddeutsche Allemeine Zeitung* newspaper that it was in sympathy with Irish nationalism, and would help make that cause a success. On 2 December, he travelled to Limburg to make his pitch to the nearly 2,500 Irish Catholic prisoners of war whom he expected to meet.

Problems began immediately. Not only were there nowhere near 2,500 potential recruits in Limburg, but the Germans had not discriminated by religion, and so there were Protestant Irishmen there as well as Catholic, along with Scottish and English prisoners who had volunteered hoping for better treatment from the Germans. After explaining to the men that Home Rule was an English trick, and the Irish in America were behind them, Casement managed to get just two volunteers.

While Casement wrangled more German logistical support for his fledgling brigade out of a sceptical Bethmann-Hollweg, Blinker Hall went to work back in Room 40. There was no direct communication between Germany and the Irish Volunteers in Ireland, so Hall was intercepting communications sent to the German embassy in Washington DC and then transmitted back to Ireland.

In December 1914 he learned that a Danish ship had been commissioned by the Germans to take Casement to Ireland. Hall, proving his mettle at the practicalities of intrigue, chartered a yacht, the appositely named *Sayonara*, and crewed it with a group of sailors faking American accents and Irish republican politics. The yacht was commanded by a Royal Navy officer, and owned by one Colonel MacBride of Los Angeles, an Irish-American pro-German, who was in reality Major W. R. Howells, a Special Intelligence Service officer. Hall's intelligence was inaccurate, and once again Casement eluded him, but the *Sayonara* aimed to draw out republican sympathisers as it put into ports along the Irish coast, its mission achieving unexpected realism due to the fact that Royal Navy patrols, unaware of its true purpose, harassed it off the west coast of Ireland.

*Members of Casement's Irish Brigade in Germany, 1915*

Meanwhile, Robert Monteith, a former British soldier who had served in the Boer War and was deported from Ireland in 1914 under the Defence of the Realm Act, was dispatched by Clan na Gael to help Casement whip his crew of freedom-fighting soldiers into shape. Casement had only managed to recruit 55 men, for despite his talents as a human rights crusader, he didn't speak the common-man language of the soldiers, and could not convince them that by supporting his brigade they were not, in fact, fighting for Germany, nor against their Irish kin serving in the British army.

The Germans were stretched on two fronts in Europe and would not give Casement more of their own soldiers, so he appealed to Clan na Gael to send Irish-American reinforcements – which they did not. Monteith, however, had procured new uniforms for the men and, astonishingly, had persuaded the Germans to let them have machine-gun training. Casement was snobbish about Monteith, who, after being discharged from a 'real' army in the ranks, had now been awarded the Irish Volunteer rank of captain. He wanted someone of higher legitimate officer stature to lead his 'Army of Deliverance', but he was stuck with Monteith.

Clan na Gael did send money, but as Casement's plan wavered, changing from an invasion of Ireland to fighting the British in Egypt, back in New York the old Irish patriot John Devoy was losing patience. With the threat of conscription looming in Britain, Devoy saw that his force of Irish Volunteers could soon be drafted into the British army, or arrested. On 10 February 1916, Blinker Hall's team intercepted a golden piece of intelligence: a decrypted message from John Devoy delivered to the German embassy in Washington for transmission to Berlin, telling the Germans unequivocally what the Irish wanted:

Unanimous opinion that action cannot be postponed much longer. Delay disadvantageous to us. We can now put up an effective fight. Our enemies cannot allow us much more time. The arrest of our leaders would hamper us severely. Initiative on our part is necessary. The Irish regiments which are in sympathy with us are being gradually replaced by English regiments. We have therefore decided to begin action on Easter Saturday. Unless entirely new circumstances arise we must have your arms and munitions in Limerick between Good Friday and Easter Saturday. We expect German help immediately after beginning action. We might be compelled to begin earlier.

Three weeks later, on 4 March, Room 40 intercepted another message, from Berlin to the German embassy in Washington, which detailed a bold plan:

Between April 20–23, in the evening, two or three steam trawlers could land 20,000 rifles and ten machine-guns, with ammunition and explosives at Fenit Pier in Tralee Bay. Irish pilot-boat to await the trawlers at dusk, north of the island of Inishtooskert at the entrance of Tralee Bay, and show two green lights close to each other at short intervals. Please wire whether the necessary arrangements in Ireland can be made secretly through Devoy. Success can only be assured by the most vigorous efforts.

Subsequent interception of communication between Berlin and Washington by the British later that month gave them a full picture of the rebellion that Casement had planned. The one thing they did not know was that Casement had changed his mind. With 20,000 guns and no Irish or Germans to carry them, he believed that the rebellion was doomed. He would go to Ireland and convince his colleagues in person to call it off.

He did not convey his revised mission to the Germans as he boarded the submarine U-20 in Wilhelmshaven on 12 April, accompanied by Robert Monteith and Daniel Julian Bailey, a POW recruit to the Irish Brigade. The three men transferred to submarine U-19 due to technical woes after a day and a half at sea, and at midnight on the 20th they arrived at the entrance to Tralee Bay, to rendezvous with the *Aud*, a captured British ship disguised as a Norwegian steamer, commanded and crewed by German sailors wearing civilian clothes. In addition to 20,000 Italian-made rifles, the *Aud* carried 10 million rounds of ammunition, 10 machine guns and a million rounds of machine-gun ammunition, as well as explosives, landmines, bombs and hand grenades.

The *Aud*, however, was not there. The skipper had made navigational errors and the ship was several miles away, shadowed by the Royal Navy. HMS *Bluebell*'s captain challenged her, and, unsurprisingly, was sceptical of her claim to be en route to Genoa from Bergen, in Norway, and taking bearings off the west coast of Ireland. He ordered the ship to put in at Queenstown for inspection. About a mile out of port, the *Aud* stopped, raised two German naval ensigns and evacuated her sailors – now in German naval uniform – into lifeboats. There was an explosion, and the *Aud* sank to the bottom of the bay.

The pilot boat that was supposed to meet the submarine failed to materialise, and as the commander of U-19 didn't want to go too close to shore, Casement and his two colleagues floated in from the U-boat on an inflatable dinghy. The Royal Irish Constabulary had alerted coastal dwellers to watch for anything suspicious along the waterfront. At 4 a.m., Mary Gorman, who was milking her cows, spotted the three men landing on the beach, soaked to the skin after their dinghy had overturned. She raised the alarm.

Casement sent Monteith and Bailey off to find help in the countryside. Bailey was arrested; Monteith evaded capture and eventually made it back to the United States. Casement himself was found 'sheltering in an old ruin called M'kenna's Fort, where, on being arrested, he gave [as his] name [that] of a friend with whom he used to stay in England'.

Basil Thomson, head of Special Branch, happened to be on Zeppelin duty at New Scotland Yard that Saturday night, when the phone rang at 10.30 p.m. 'You know that stranger who arrived in the collapsible boat at Currahane? Do you know who he is?' asked the unidentified caller on the other end. Thomson knew well the caller's voice, and thought he was joking. '"I am not," replied the caller, "and he will be over early tomorrow morning for you to take him in hand." It was not necessary for either of us to give a name. We had been expecting Casement's arrival for many weeks.'

So while the armed Easter Sunday uprising that Roger Casement had planned, reconsidered, and then failed to stop was preparing to launch its attack on the British in Ireland, Casement was being interrogated by Basil Thomson in London. 'He walked into the room rather theatrically – a tall, thin, cadaverous man with thick black hair turning grey, a pointed beard, and thin, nervous hands, mahogany-coloured from long tropical service,' Thomson recalled. Once Casement had admitted to high treason, and the shorthand reporter had been dismissed, he opened up, telling the Special Branch boss that while the Germans had wanted him to go to Ireland on the gun-running ship, he'd insisted on travelling by submarine 'to warn the rebels that they had no chance of success ... He was very insistent that the news of his capture should be published, as it would prevent bloodshed.'

On Easter Monday, 24 April, the Irish rebels launched their attack, which ended the following Saturday with their surrender, despite having lost only one of their captured positions to the British army. Blood had indeed been shed: 466 people had been killed, with more than 2,000 wounded. The leaders of the rebellion were swiftly executed, martyrs to the Irish cause and immortalised in republican history. It was not the glorious victory imagined by Casement's Army of Deliverance, but the Irish republic had bloodily emerged into the light of political day.

Sir Roger Casement, with his knighthood, his international celebrity and his powerful British and American friends, was not the type of case for a rough-justice military tribunal. His trial in London in June 1916 was prosecuted by no less a light than the British government's highest-ranking law officer, Attorney General Sir F. E. Smith, who two years earlier had been one of the loudest Tory advocates of resistance in Ulster to Home Rule, and as such was seen by Casement and many others as a traitor himself.

Even so, his prosecution of Casement was, according to Casement's Irish lawyer, Serjeant Sullivan (no Englishman would represent him), 'chivalrous and generous to the weaker side'. Indeed, it was Sullivan who ensured Casement would be executed, despite appeals for clemency from the cream of the Anglo-Irish literary establishment: W. B. Yeats, Arthur Conan Doyle, George Bernard Shaw, and G. K. Chesterton among others. Sullivan refused to use Casement's so-called 'Black Diaries' in his defence as proof of insanity, diaries dating from 1903 that had been found earlier in

*Sir Roger Casement during his trial in London 1916*

1916 in a trunk Casement had forgotten about in his London flat. As Basil Thomson later wrote, expressing a sentiment of a time when homosexuality was still a half-century away from decriminalisation in Britain, 'It is enough to say of the diaries that they could not be printed in any age or in any language.' Thomson and Blinker Hall made sure that the most extreme passages circulated among influential members of the government and media in order to damn Casement to death by scuppering any sympathy for the 'statesman'.

The diaries contained explicit accounts of Casement's considerable homosexual encounters at home and abroad, with an obsessive notation of the genital size of his partners. Indeed, a sample from 14 December 1911 in Manaus, Brazil, would suggest that Casement was either freakishly superhuman or a fabulist:

Out to Joao Pensador by tram, bathed there and walked back to cricket and to the bathing pool there. Seven school boys (one a cafuzo [of mixed parentage] 17–18) and 5 of them white and 4 had huge ones and all pulled and skinned and half cock all time. One a lad of 17 had a beauty. All 'gentlemen'. After dinner out at 7.10 met Aprigio on seat. Stiff as poker and huge. So together to terracos baldia [uncultivated terraces] where sucked and then he in. Left and met Antonio my sweet Caboclo of last time and he followed and showed place and in too – hard. Huge testeminhos ['witnesses', i.e., testicles] and loved and kissed. Nice boy. Then young Alfandega Guarda mor [chief customs inspector] darkie big and nice – bayonet and felt it huge and stiff as his bayonet. Awfully warm. Nice lovely Italian boy passed at 11 and smiled and so to bed.

Such diaries could be used as proof of insanity due to their fantastic nature; any competent counsel could have argued, invoking the morality of the time, that Casement couldn't possibly have had that many homosexual encounters, with that many superlatively well-endowed partners, and so in fact was mentally unhinged. Nevertheless, Serjeant Sullivan not only refused to read them, he refused to use them, as he wanted Casement to plead not guilty. In an interview four decades later he explained, 'I finally decided that death was better than besmirching and dishonour.'

In other words, it would have been impossible for Casement to be lionised as an Irish patriot once portions of the diaries were published to justify the leniency. The lurid diaries, combined with the discovery of a document proving that Casement planned to use his Irish Volunteers to fight the English in Egypt, were enough to remove support from even the most fervent apostles of clemency. And so, after being received into the Roman Catholic Church and thus completing his Irish circle, Casement was hanged on 3 August 1916 at Pentonville Prison. Thanks to Room 40, Britain had narrowly avoided full-scale Irish civil war. But the war in Europe and the Middle East was only getting worse.

# Chapter 12

# TESTING TIMES

1916 saw the British armed forces enter Hell. On the Western Front, the Battle of the Somme was the largest land battle ever seen, while in the North Sea the Battle of Jutland was, while much shorter in duration, the naval equivalent. These titanic military encounters were matched on the home front by unprecedented mobilisation of men and resources as the state took complete control of every aspect of its citizens' lives. Conscription was introduced and women were entering the labour force, taking jobs previously done by men, in ever-increasing numbers.

To meet the scale of the challenges ahead, Room 40 needed to expand its staff. During 1916, Hall added a couple of lords, a scientist, three Foreign Office men – one of whom was 'chiefly remarkable for his spats' – several writers and actors, three professors of German, a number of classicists, some City men, a caricaturist, and a handful of wounded or disabled army officers who collected the material arriving by pneumatic tube, increasing the team to around 40 members.

A sense of what the atmosphere was like in this thriving and busy little community was provided by Captain William 'Bubbles' James, who'd served under Hall on the *Queen Mary* and who now replaced Captain Hope, Room 40's previous naval expert. A traditionalist, James insisted the codebreakers wear uniform, and was less collegiate than his predecessor. Nevertheless, he brimmed with enthusiasm and was immensely proud of his men, a sentiment that shone through in his recollections of Room 40 life, which he described as 'vibrating with excitement, expectation, urgency, friendship and high spirits', filled with people 'all talking a strange language and doing strange things'.

For Dilly, a welcome addition to the Room 40 team was his good friend Frank Birch. Born in 1889 and educated at Eton, Birch met Dilly when he was an undergraduate at Cambridge studying history. A gifted conversationalist and actor, he proceeded to become a don but yearned for a theatrical career. Not long after the war, he abandoned his life as a historian and took to the stage, gaining fame for his portrayal of the Widow Twankey in the pantomime *Aladdin*, a part that led to minor roles in a string of films.

Before joining Room 40, where he was in charge of collecting and collating information, Birch was in the Royal Naval Volunteer Reserve and saw active service in the Channel, the Atlantic and the Dardanelles. He and Dilly shared a house in Chelsea, where Birch held musical parties every weekend featuring the world-famous cellist Madame Suggia. Dilly avoided these soirées by signing up for the night shift.

The rhythm and mood of Room 40 was set by the Germans' habit of changing their cipher keys every 24 hours at midnight. The job of deciphering the new key therefore fell to the night watch; their success or failure would set the tone for the day. According to James, 'a visitor entering the watch keeper's room in 1916 would either have seen three or four men, very tired and drawn, who had for the last eight hours been straining their brains to discern the cipher key ... and had been defeated, or would have seen the same men looking very cheerful and waiting to tell the relief ... that they had nothing to worry about.'

Despite working at a heightened level of efficiency, delivering information at almost real-time speed, Room 40's operational usefulness was still being treated with scepticism by naval staff, while the procedural protocols that inhibited rapid and coherent processing of its intelligence remained stubbornly in place. During 1915, when the German High Seas Fleet was relatively idle, these potentially lethal fault lines were not exposed to undue stress; in 1916, there would be nowhere to hide.

In the spring of that year, Admiral Reinhard Scheer took control of the High Seas Fleet. A firm believer in submarine warfare, he was determined

*Admiral Scheer, head of the German High Seas Fleet
during Jutland and the U-boat campaign*

to confront the British using the same tactics that had previously failed
because of his predecessor's caution. An attack on the coastal town of
Lowestoft was attempted in late April but abandoned when Scheer got a
warning from his codebreakers at Neumünster that the British Grand Fleet
was on the move.

Scheer's next major outing was scheduled for the last weeks of May.
Delayed by bad weather, he finally set sail on the 31st. Room 40 had been
alerted as early as the 17th that something was up and on the 30th produced
definitive evidence that Admiral Hipper's battle cruisers were about to leave
their bases. Jellicoe was duly informed, and with Beatty scouting ahead,
he put to sea that evening, not yet certain whether Hipper was alone or
accompanied by Scheer and the rest of the High Seas Fleet.

Given the importance of establishing Scheer's position, it was particularly
unfortunate that the task fell to Captain Thomas Jackson, director of the

Operations Division. Jackson, who was making only his third visit to Room 40, had nothing but contempt for the codebreakers: 'these chaps couldn't possibly understand all the implications of intercepted signals'. Rather than asking directly about Scheer's whereabouts, he enquired instead about the location of his flagship, denoted by the wireless call sign 'DK'. Room 40 dutifully supplied the answer: intercepted signals placed the call sign 'DK' in the Jade Bight, home of the High Seas Fleet. Satisfied that this provided confirmation that Scheer had not left port, Jackson passed on the news to the Admiralty, who then informed Jellicoe. The problem was that Scheer, hoping to outwit the British DF stations, habitually abandoned the 'DK' call sign when he set sail, assigning it to a local onshore outpost instead.

Room 40 was well aware of this routine deception, and had Jackson either been clearer about the purpose of his question or bothered to make further enquiries, the codebreakers would have revealed the real state of affairs: the placement of the call sign proved that Scheer was not staying put. Had Jellicoe known that the High Seas Fleet was behind Hipper, he would have been able to act accordingly: either by laying a trap for the enemy or by disengaging altogether. As it was, when he did discover the startling truth, it dented his confidence in any other intelligence coming in from the Admiralty over the course of the battle.

Ironically, the Germans were equally convinced that Jellicoe and the Grand Fleet were not heading in their direction either, despite getting word from Neumünster that their codebreakers had intercepted British wireless messages suggesting otherwise. In his account of the battle, Scheer claimed that he disregarded these reports because they 'gave no enlightenment as to the enemy's purpose'. His decision was supported by the German official history of the naval war: 'this intelligence, therefore, in no way affected the projected plan. On the contrary, it only increased the hope that it would be possible to bring part of the enemy's fleet to action.'

So it was that the two largest fleets ever assembled were, without realising it, on a collision course. The British force easily outnumbered the Germans, with 28 dreadnoughts against their 16 – in addition to six pre-dreadnaughts and 113 other craft – light cruisers, battleships and torpedo-armed destroyers – against their 72. As for battle cruisers, Beatty had nine to Hipper's five.

Between 3.45 and 4.40 p.m., Beatty and Hipper clashed violently, the German getting the better of an intense gunnery duel, losing only one battle cruiser while destroying three of Beatty's. To see a warship go down was a harrowing and unforgettable experience. When the end arrived, these great nautical beasts would disappear beneath the waves in a matter of minutes. One German commander remembered the demise of the *Queen Mary*: 'black debris of the ship flew into the air … gigantic clouds of smoke rose, the masts collapsed inwards … finally nothing but a thick black cloud of smoke remained where the ship had been.'

For the few survivors plunged into the freezing North Sea, conditions were nightmarish. A midshipman recalled how 'the surface of the water was simply covered with oil fuel which tasted and smelt horrible. I smothered myself all over with it, which I think really saved my life as the water was really frightfully cold.'

Meanwhile, on the severely scarred ships limping away from the fight, terrible fires were raging, their venomous and vengeful flames inflicting horrific suffering: 'some of the dead were so burned and mutilated as to be unrecognisable. The living badly burned cases were almost encased in wrappings of cotton wool and bandages with just slits for eyes … a grim, weird and ghoulish sight.'

Having come off worse, Beatty pulled away and began to draw Hipper towards Jellicoe, still unaware that Scheer was close by. Suddenly the High Seas Fleet came into view. Beatty signalled their presence to a shocked Jellicoe, who immediately began to assemble the Grand Fleet into a fighting formation. Still unaware of what lay ahead, Scheer kept up his pursuit of Beatty until, at 5.59 p.m., he spotted Jellicoe. It was too late for evasive action. The moment that both sides had done their best to avoid had finally come.

The power of Jellicoe's guns was too much for Scheer to bear: to stay and trade blows was suicidal. He ordered his fleet to veer away sharply, then, in a moment of madness, turned it back towards the British dreadnoughts before realising his error and signalling another retreat, this time for good. To cover his escape, Scheer sent his destroyers directly at the British, hoping their torpedoes would delay Jellicoe. Their appearance was enough to make Jellicoe hesitate, and with darkness falling he instructed his ships to pull back rather than risk destruction by Scheer's underwater missiles.

Scheer had escaped a severe mauling but was still a long way from home. Not wanting to face Jellicoe again at dawn, he opted for the most direct way back to port, which entailed literally passing in front of the Grand Fleet. Jellicoe, keen to avoid a night battle, and ignorant of his opponent's intentions, decided to put his fleet in a position to intercept Scheer come daylight. Even though there was sporadic fighting as his scouting squadrons ran into Scheer's rearguard, Jellicoe did not realise how close he was to the rest of the German force.

He would have done if the stream of information coming out of Room 40 had been properly handled. At 10.41 p.m., the Admiralty forwarded a message to Jellicoe: 'German battle fleet ordered home ... course south-south-east.' After the Jackson debacle, and having received another message shortly before this one that he knew to be inaccurate – the error was down to the German commander who sent it rather than Room 40 – Jellicoe, who felt that 'experience earlier in the day had shown one might be misled if trusting too much on intercepts', chose to ignore it.

Then a decode that included Scheer's request for Zeppelin reconnaissance at dawn near the German coast – which, as Jellicoe later observed, 'was practically a certain indication of his route' – was not even sent to him. Between 10.43 p.m. and 1 a.m., seven other messages never got beyond the Operations Division. Jellicoe was scathing in his criticism: 'the lamentable part of the whole business is that, had the Admiralty sent all the information which they acquired ... there would have been little or no doubt in my mind as to the route by which Scheer intended to return'.

No one ever claimed responsibility for this unforgivable failure in communications. The rumour was that Henry Oliver, the conduit for Room 40 intelligence,, who was often on duty 24 hours a day, had crept away to catch a few hours' sleep and left only a junior officer in charge.

As a result, Scheer made it back to port without further serious loss. Afterwards, both sides claimed victory. Certainly the Germans had reason to be triumphant: they'd lost 11 ships but had sunk 14, had suffered 2,551 dead and 507 wounded yet killed 6,094 British seamen and wounded another 674, while taking 177 prisoners. Yet the overall strategic position had not changed. The British navy's pride and aura of invincibility had been dented, but its strength was not diminished enough to alter the balance of power.

This was cold comfort to the Admiralty as it faced a barrage of criticism. Jellicoe was accused of excessive timidity and replaced by the more cavalier Beatty. Detailed examination of the course of the battle led to the obvious conclusion that the current method of disseminating Room 40's product was not fit for purpose.

Blinker Hall, who had long resented the Operation Division's hold over Room 40, was now able to wrest control away from it. At last he could realise his ambition to turn Room 40 into a fully fledged and properly integrated intelligence organisation supplying its own daily reports. Under this new regime it would go from strength to strength, aided by the fact that the Germans continued to reject any suggestion that their codes were compromised, even though, once again, the British had miraculously been in the right place at the right time.

Scheer, bloodied but not beaten, shifted the focus of his strategic thinking to his beloved U-boats. He believed that 'a victorious end to the war … can only be looked for by the crushing of English economic life through U-boat action against English commerce'. The outcome of the war at sea would be decided in the Atlantic.

While wireless communications were fundamental to the war at sea, on the Western Front the British were reluctant to embrace the new technology. Unlike the navy, the army was yet to be convinced of its usefulness. Initially, the chiefs of staff had been intrigued by radio, but when Marconi's magic boxes were test-run during the Boer War, the results were disappointing. Though it was the engineers who operated them who were at fault, the army blamed the equipment. Experiments with radio did continue, but on a small scale. In 1905, the first Wireless Section was formed. By 1914, their number had risen to a mere ten, and the portable sets they used were extremely unreliable. According to the Royal Engineers' account of the conflict, *The Signal Service in the European War of 1914–1918* (1921), wireless was 'looked upon with suspicion and dislike'. The generals preferred the telephone.

Unfortunately for them, in early 1915 the Germans developed an ingenious listening device that could eavesdrop on Allied phone activity. The Moritz set had the capacity to tap into telephone signals over a 3,000-yard radius by accessing the earthed current that relayed British messages: copper plates buried in the soil picked up the low voltage generated by these signals and Moritz translated them back into their original form.

Operators were sent to Berlin for a six-week training course that included lessons on British slang expressions. For extra protection, the Moritz sets were kept in specially reinforced dugouts. They were under strict instructions not to let any fall into enemy hands, while soldiers were ordered not to say a word about the device if they happened to be taken prisoner.

*Field set trench telephone*

The Moritz sets fed on a rich diet provided by the careless and sloppy use of telephones by the British – though the French weren't much better. As neither had developed equivalent technology yet, they operated in blissful ignorance of the German capability, nattering away *en clair*, or when using code, using it badly.

In the major offensives launched by the Allies, the elusive goal was 'breakthrough' – getting beyond the ranks of German defensive lines and into open country – where the cavalry, not yet abandoned as a potentially decisive weapon, could run riot. To achieve this, the first days of battle, before the enemy could rally, pour in extra troops and launch counter-attacks, were critical. Surprise was essential. However, the Moritz sets meant that the Germans were forewarned and forearmed. Their defensive capability gained immeasurably from insights into who was facing them, in what numbers and where. The exact locations of the series of assaults launched by the French during 1915 were known well in advance.

Major R. E. Priestley, author of the Royal Engineers history of the Signal service, wrote that 'the question of enemy overhearing had arisen in the summer of 1915'. A disturbing pattern was emerging: carefully planned raids and minor attacks were 'met with hostile fire, exactly directed and timed to the minute'. This was happening too often to be mere coincidence. All the indications suggested that the leakage of information was 'intimately connected to the extension of telephone to the front-line trenches'.

However, the warning signs were ignored. In his diary entry for 4 April 1915, Brigadier General John Charteris, Haig's intelligence supremo, the man responsible for keeping the Commander in Chief of British forces up to speed, noted in his diary a conversation he had with the commander of the 9th Battalion Highland Light Infantry (Glasgow Highlanders), who thought the enemy had 'very good information on our front-line dispositions' based on the fact that when his men approached the German trenches they were greeted by 'a good imitation of a Glasgow tram conductor's voice' and a gramophone playing 'Stop Your Tickling Jock'.

Though Charteris found this amusing, he also saw the serious side: clearly the Germans had some source of information, most likely captured

British soldiers. It did not even occur to him that there might be another explanation or that the incident was alarming enough to merit further investigation. This failure to join the dots was to have even more serious consequences – so serious that they finally alerted GHQ to the problem – during the Battle of the Somme.

The British army was not meant to lead a major offensive in 1916; that dubious honour had fallen to the French. The massive German onslaught on Verdun, designed to bleed the French army to death, necessitated a change in plan. The onus was now on the British to take the initiative. Haig pondered his options and settled on the Somme. This five-month campaign, the biggest battle of the war so far, would consume hundreds of thousands of lives for the gain of a few miles; it has become synonymous with everything mind-numbingly awful about the Western Front. Had telephone security been better, things might have turned out differently.

The British plan relied on achieving major gains within 24 hours of the infantry going over the top. Intelligence derived from Moritz gave the Germans a clear picture of what to expect; what they couldn't discover was when the assault would come and when the immense, relentless artillery barrage that preceded it would end. Day after day they prayed for the guns to fall silent as hell rained down on them.

A few hours before dawn on 1 July, as the Tommies readied themselves in their forward trenches, the Germans were gifted the information they needed: a Moritz station intercepted a message from British HQ offering encouragement to the troops waiting to advance. Any possibility that the Germans would be taken by surprise was gone, and with it the chance of a decisive outcome.

The first day on the Somme was a disaster as yet unparalleled in the history of the British army. Reflecting on the reasons for this unholy mess, commanders finally woke up to the problem of loose talk. A memo issued by the General Staff on 23 July noted that there was clear evidence that a 'German system of overhearing' was being extensively used, and that 'it must be assumed that the enemy has listening apparatus'.

Incontrovertible proof that these suspicions were well founded came during July, when the German stronghold at Ovillers-la-Boisselle was captured after numerous costly attacks. The occupying troops discovered that the Germans had left behind a complete copy of the operational orders of the British corps that had repeatedly tried to storm the position.

Stringent efforts were made to enforce greater discipline. A lengthy memo appeared in October on 'The Indiscreet Use of Telephones'. It demanded an end to unnecessary gossip, and the introduction of silent hours, and warned that any disobedience of those orders would be severely punished. It was too little, too late. The Moritz sets were still able to identify 70 per cent of the British units deployed over the course of the battle, knowledge that was used to devastating effect.

One of the most revealing accounts of life as an intelligence officer on the Western Front was written by Ferdinand Tuohy, a journalist who'd worked for the Northcliffe press. In it he constantly drew attention to the lack of imagination displayed by his superiors and their tendency to be reactive rather than proactive. That he was eventually to pioneer a wireless interception and codebreaking system that would help the British learn something useful from the Somme campaign is testament to his persistence and his refusal to play by the rules.

Tuohy began the war reporting for *The Times* and covered the First Battle of Ypres. In January 1915 he was sent to Poland to investigate the situation there; stuck in Warsaw, and unable to get to the front, he 'had no option but to continue writing the stereotypical stuff' about 'fearless and faithful Ivan', though in reality he found the Russians 'boorish and primitive'. Nevertheless, he pitied the peasant conscripts who formed the majority of the army; they were 'massacred like mutton', had no leave, no pay and no mail and, due to 80 per cent illiteracy, were 'unable even to break, by reading or writing, the desperate boredom of trench life'.

Eventually Tuohy got amongst the action, finding himself at the Second Battle of Bzura, fought along a 30-mile front that was 'bleak, freezing, desperate'. Signing on with an ambulance unit, he then moved on to the great Austro-Hungarian fortress Przemyśl, which guarded the gateway into the Carpathian Mountains and had fallen to the Russians in March 1915 after a bitter siege. He witnessed scenes of appalling devastation, 'ruined, diseased and starving villages all around'. By the end of May, the citadel was back in enemy hands and Tuohy was back in England recounting what he'd seen to a shocked Northcliffe, who was appalled that 'we know practically nothing of all this ... and I don't believe the cabinet does either'.

Decent information about what was actually happening on the Eastern Front, where vast armies ranged over hundreds and hundreds of square miles, retreating one minute, advancing the next, was extremely difficult to come by, with even commanders on the ground finding it hard to keep track of the fighting. So Northcliffe dispatched Tuohy to the Foreign Office to fill them in. They directed him to the War Office, where he was asked to join the Intelligence Corps. After a few months studying maps and being taught about the German army, he was sent to Ypres.

Without listening sets of their own, officers like Tuohy were obliged to crawl up to the German lines under cover of darkness to try and overhear enemy chat. This rudimentary approach finally ended after the French developed a listening set comparable to the Moritz, which the British called ITOC. It could pick up German conversation some 500 yards from the front line. The sets were located in a dugout with wires leading out of it into no-man's-land and operated by two interceptors, young men of 'the clerical breed, experts in dialectical German ... with telephone headpieces glued to their ears ... pencil in hand' ready to jot down any stray snatches of conversation. Tuohy would sift through the results, picking out from the 'voluminous twaddle' any useful titbits relating to matters of tactical concern.

The fact was, German telephone procedures, established early and based on the belief that if they had listening devices then so must the Allies, were far more secure. Special codes, consisting of simple word substitutions, were introduced to reduce the risks of interception: 'snake' for 'casualty', 'monkey' for 'prisoner', 'carp' for 'Frenchman', 'dried cod' for 'British soldier', 'walrus' for 'Russian', etc., giving rise to sentences like 'we had some snakes, but brought in several monkeys including a walrus, a carp and a dried cod'.

Tuohy realised that 'the enemy was obviously getting more vital results from Moritz than we had ever got from ITOC'. Where the ITOC sets came into their own was as 'an instrument for policing the conversations of our own men'. What they revealed was that old habits died hard: too much pertinent information was still being bantered about. Over the course of one month, a single set heard 40 units referred to by name and talk relating to the movement of troops, operation orders and positions behind the line.

At least now there was a way to keep track of the leaks, much to the resentment of the men, who felt spied on by their own side. The result was that 'in due course a distinct hush fell over our front trench system'. This hush spread to the German lines, and by the end of 1917 there was precious little for the listening sets to overhear. By then Tuohy had played a vital part in establishing an interception system that for once was ahead of the Germans.

While the daring, blind courage and glamour of the ace fighter pilots engaged in aerial combat grabbed the headlines, the real tactical value of the air force came from reconnaissance. Aside from the quantum leaps in photographic technology that produced ever more detailed images of the battle zones, wireless was employed by the spotter planes to register targets for the artillery and monitor events on the ground.

In early 1915, the Royal Flying Corps, working in tandem with the Marconi Experimental Laboratory in Surrey, developed a lightweight transmitting set with a range of 20 miles. Pilots were taught by Marconi engineers how to operate it, and by 1918 this primitive radio was installed in 600 planes. Messages from the pilots to the ground were encoded using a clock-face system, with different segments relating to particular map references. Armed with this technology, flying low over the German lines, these planes greatly improved the accuracy and tactical efficiency of the artillery: counter-battery fire – shelling the enemy's guns – relied almost exclusively on information coming from the air.

The Germans had wireless in their planes too, and the British were soon intercepting messages sent by them. However, analysis of this material progressed slowly until Tuohy got involved. He described how, in the late summer of 1915, 'a wireless set with a battery near the ramparts of Ypres' was picking up the call sign of the enemy spotters, 'a high pitched, quivering note', as they communicated the position of British troops to their artillery. Tuohy quickly understood that if you could 'connect given German wireless call sign with known ... hostile batteries', the British guns could then turn their sights on the German artillery that had been identified by the planes' wireless signals 'and crump its personnel back into the bowels of the earth'. Equally, the position of the German spotter planes could be tracked, leaving them vulnerable to the attentions of the British fighter squadrons.

Armed with a mass of material, Tuohy attempted to break the codes used by the German planes. Sitting up all night 'bent over fragmentary wireless hieroglyphics', he laboured for months 'trying every conceivable juggling of lettering based on German colloquialisms' until, after much guesswork and deduction, he solved their letter code. A parallel numerical code continued to defy him until a downed plane was recovered with its wireless set and code book intact.

By early 1916, the British army had what it needed to gain a real advantage. But it failed to act: 'GHQ got everything we had, yet we got little in return.' Tuohy put the unforgivable delay down to the intractability of the system: 'minds were warped by a departmental outlook' and 'mountainous documentary files came into being in which the vital end to be achieved was lost, submerged'. Meanwhile, 'British soldiers were dying in thousands as a direct result'.

Six months later, and only because the desperate situation on the Somme demanded it, this system of analysing messages sent by German planes was finally given its due. Every evening Tuohy briefed General Hugh Trenchard, head of the RFC, on the aeroplane wireless activity of that day.

The impact of these measures was felt immediately by the German army. As the number of prisoners and abandoned dead and wounded mounted, so did the volume of letters, diaries and documents acquired by British intelligence. A common theme was the ever-present threat posed by the RFC. One soldier complained that 'the English are always flying over our lines, directing artillery fire, consequently getting all their shells … right into our trenches'. Another noted in his diary that 'once a battery has been located it can result in it attracting 2,000–3,000 shells', while an officer admitted that 'whenever the slightest movement was visible in our trenches … a heavy bombardment of that section took place'.

By the time Tuohy left France for pastures new, he was delighted to see 'a network of 14 wireless intelligence posts established'. By the end of the war there were a thousand, manned by 18,000 staff. Between October 1916 and March 1917, they successfully decoded messages from more than two-thirds of the flights made by enemy spotter planes. During that year, 50–60 per cent of the Germans' infantry divisions and artillery formations were pinpointed thanks to the methods advanced by Tuohy.

Yet despite all the effort and innovation, coupled with the undoubted improvements that wireless intelligence brought with it, victory remained an ever-receding prospect. The blood of nations was draining into the mud, yet the staggering casualties suffered by all sides did not appear to be bringing the end any nearer.

# Chapter 13

# THE FAR WESTERN FRONT: AMERICA

Given that all the combatants were under tremendous strain and getting ever closer to breaking point, the role of America became increasingly crucial to the outcome of the war. It had always been Germany's contention that by giving the Allies so much financial and material support, America was effectively a belligerent power, a status that justified action against it. The secret, and often not so secret, war that Germany was waging on US soil would reach a terrifying climax thanks to the machinery of terror constructed by Franz von Rintelen, the self-styled Dark Invader.

Before his unscheduled departure from America and subsequent arrest by Blinker Hall, von Rintelen had laid the groundwork for a campaign of death and destruction that included bomb factories, ingenious explosive devices and, ultimately, biochemical weapons. Though Room 40 knew about von Rintelen's subversive actvities in the USA, the most spectacular and destrictive attack on American soil would also display that no matter how far the reach of Room 40 went in fighting the intelligence war, Blinker Hall's team of codebreakers could not prevent acts of sabotage against the United States for which they had no intelligence – in part a function of a large number of isolated German cells communicating internally, and the absence of any kind of effective national security force in the USA.

Von Rintelen's sabotage network was nothing if not ambitious. One destructively simple invention came to him courtesy of Robert Fay. When the war began, the 33-year-old Fay had been called into action as a lieutenant in a German infantry battalion that saw heavy action in the Vosges mountains and Champagne. As a mechanical engineer, Fay took special interest in the quality of the Allied shells that were trying to kill him. He devised a way to stop them at source, and with $4,000 and a neutral passport from Military Intelligence, Sektion IIIb, he sailed for New York City in April 1915, arriving there shortly after von Rintelen.

Fay's invention was simple and effective: he had designed a self-detonating bomb to destroy the rudders of ships sailing to Europe with supplies for the Allies. Von Rintelen dispatched Fay and some of his own sea captain confederates to buy a well-hidden plot of land far from nosy neighbours where they could test a prototype. The sailors built the stern of a ship out of wood, and attached an actual rudder. Fay applied a detonator, on the tip of which was a needle-nosed pin connected to the rudder shaft. The idea was that as the shaft turned, the pin turned with it, with the sharp end boring into the detonator until it made contact with the explosive, and blew up the rudder.

*Robert Fay, the designer of the ingenious 'rudder bomb'*

Fay demonstrated his prototype to the sea captains, observing from a respectful distance, but after an hour of turning the rudder, nothing had happened. Then suddenly fragments from the wooden stern of the ship were flying at the captains, and Fay himself was flying up in the air, landing hard and injuring his ribs. Trees were blown away and the assembled spectators had to put out a fire, after which, as von Rintelen drily related, 'they then got into the car and returned to New York to report to me that the invention had functioned efficiently'.

Fay drove a motorboat into New York Harbor to attach his first set of rudder bombs. After a spot of engine trouble, which he managed to repair, he completed his mission and awaited the results. Shortly afterward, two ships mysteriously lost their rudders at sea. Fay was now in the rudder bomb business. However, his success, and the concern it had generated in the media, meant that his subsequent sabotage had to be more covert. His solution was to mount his rudder bombs on cork platforms and swim them out to their targets under cover of darkness. More success followed, and von Rintelen franchised the rudder bomb to other crews along the eastern seaboard.

In May 1915, the same month in which the *Lusitania* was sunk, a German naval officer visited the bomb-making factory in Hoboken, New Jersey, of one of von Rintelen's key collaborators, Dr Walter Scheele. The visitor that day, Erich von Steinmetz, had brought with him a powerful weapon inside a suitcase, and along with it a rollicking adventure tale with an unexpectedly feminine twist. Steinmetz had carried his weapon from Romania across Russia, and then through Siberia to Vladivostok, where he set sail for San Francisco, taking the train to New York. As he journeyed through high-testosterone war zones and twitchy checkpoints, he purchased women's dresses to disguise himself as a modest and harmless female. Once in New York, he went straight to von Papen's war office at 60 Broadway, and was duly sent to see Germany's longest-serving American spy.

Dr Walter Scheele took one look at what was inside von Steinmetz's well-travelled suitcase and knocked the visitor down with a swift punch. Despite having no qualms about bombing ships at sea, or ruining shipments of cornmeal with blue methylene dye that he concocted in his lab, Scheele's loyalty to the Fatherland, whom he had been serving as Germany's 'eyes' in America since 1883, would not entertain a venture into chemical warfare, even though his countrymen had used chlorine gas to international outrage at the Second Battle of Ypres in April and May of 1915. And von Steinmetz was, to Scheele's mind, presenting something even worse: biochemical war in the form of a culture of pathogens to poison horses destined for the Allied cause.

From pulling artillery guns, to bearing reconnaissance riders and providing transport over muddy and rough terrain, to hauling wagons laden with equipment, horses were used extensively by all combatants. The British Remount Service's largest American horse depot, in Newport News, Virginia, saw nearly 500,000 horses shipped to the Allies during the war. The American Expeditionary Force used 182,000 horses and mules during their campaign, of which more than 63,000 died. And in Germany, the horse population was reduced by 1.3 million during the war.

The importance of the horse was not in doubt to anyone, and the German High Command understood that torpedoing ships transporting men and horses was not the only way of eliminating critical Allied stock. When von Steinmetz's deadly pathogens turned out to be duds, likely due to his mishandling of them during his cross-dressing odyssey to America, they turned to a man who knew what he was doing, and who had decided to betray the land of his birth.

Anton Dilger was born in the horse country of Virginia in 1884, the tenth child of Elise, a spiritualist, and Hubert, a German hero of the US Civil War. He spent the first few years of his life in a mansion on Greenfield Ridge Farm, speaking more German than English, hearing stories of the glories of the Fatherland from his maternal grandfather, and immersing himself in the world of horses around him. When he was ten years old, his sister married a wealthy German businessman and took young Anton to live with them in Mannheim. There he stayed, becoming increasingly German as school gave way to university, where the handsome and intelligent young man studied medicine at the prestigious University of Heidelberg.

Dilger graduated as a doctor with a specialty in surgery and micro-biology, having studied the microbial origins of wound infections and how to prevent them. With the coming of war he joined the German army as a surgeon. While performing his duties at the German Red Cross hospital in Karlsruhe – just 50 miles from the Western Front – a terrible tragedy occurred that turned his mixed feelings about the Allies into implacable hatred.

On the holiday afternoon of 22 June 1916, French planes attacked the city. They had been aiming for the train station, but using outdated maps instead bombed the Hagenbeck Circus, filled with children still in their white procession robes after earlier celebrating the Feast of Corpus Christi in this predominantly Catholic city. Dilger was on hand when scores of wounded children, their white robes stained with blood, were rushed into his hospital. He had been awake for 48 hours, and the sight of dead children and the primal wailing of the mortally injured (and their parents) caused him to break down. The bombing killed 120 people, including 71 children, and injured another 169. It would come to define Dilger's mission against the country of his birth.

Once back in America, Dilger established his bio-terror factory – known as 'Tony's Lab' to his confederates – in the basement of a house in Chevy Chase, Maryland, just six miles from the White House, and there propagated anthrax and glanders. Assisted by his older brother Carl, a brewer, he concocted his poisonous cultures with the greatest of professional caution. He knew what an accidental dosage of anthrax could do to a man's lungs. He needed a careful delivery system to transit the poisons to the equine population of America without starting a mass infection in humans. And such a system had been put in place a year earlier by none other than the now imprisoned Dark Invader Franz von Rintelen, who had travelled to Baltimore to enlist soldiers in his covert war.

The port of Baltimore was an important conduit for shipping war materiel to the British and French, and it was there that von Rintelen had found two exceptionally well-connected Germans. Henry Hilken had emigrated to the US in 1866, married an American, and prospered. He was the honorary German consul in Baltimore, and local head of the

Norddeutscher Lloyd, Germany's largest shipping fleet. His son Paul, in his mid thirties, with a moustache and rakish swept-back hair, was his father's trusted lieutenant, and lived with his wife and young daughter in a big house in the exclusive Roland Park neighbourhood.

Von Rintelen found Paul Hilken eager to help his father's Fatherland (an act that would cause a lifelong breach between father and son). He would act as von Rintelen's Baltimore paymaster, and he had the perfect operative to carry out a southern version of Rintelen's cigar-bombing project.

Frederick Hinsch was a huge, blonde, hard-drinking sea dog in his mid forties, the feared captain of the SS *Neckar*, a ship in the Norddeutscher line that was interned in Baltimore by the war. A commander of men through both cunning and brawn, Hinsch soon ran a bomb-planting team of stevedores, led by Eddie Felton, an African-American dockworker who saw the $150–200 a week that he received to run his largely black crew of saboteurs as fair compensation for his so-called freedoms in the land of the free.

When Hinsch first met Dilger in his basement lab, the German sea captain was keen to ascertain that the poisons Dilger was cultivating actually worked. Dilger opened the cage where he kept infected guinea pigs and showed Hinsch that the animals were almost dead. He had cultivated deadly cultures in vials labelled #1 and #2. The #2 vial should be rubbed in horses' nostrils or poured into their feed and water troughs, while the #1 poison should be injected with a syringe.

Satisfied, Hinsch paid Dilger and left the house in Chevy Chase with his boxes of lethal vials wrapped in brown paper tied up with string. The first and most modern use of biological weapons in the USA was about to begin, as Felton and his saboteurs fanned out to infect the remount depots along the eastern seaboard of the United States, using rubber gloves and needles. Horses and mules would fall ill and die, but the campaign never succeeded in destroying horses the way that the masterminds of Sektion IIIb in Berlin had hoped – and not nearly as effectively as the war itself had done.

Dilger would return to Germany, and make one final visit to America before disappearing into Mexico under the *nom de guerre* 'Delmar'. While his campaign of bio-terror had ended in the United States (the Germans launched others in Romania, Spain, Norway and South America), his accomplices Hinsch and Hilken had money and an army, and they would launch the biggest attack inside the United States that the country had yet seen.

The networks of agents and sabotage operations established by the Germans, and particularly Franz von Rintelen – now interned by the British at Donnington Hall – exploded in New York City at the end of July 1916, when the Battle of the Somme was about to enter its second month of mass slaughter. It was an explosion that no amount of Room 40 monitoring of communications between Berlin and von Bernstorff could have revealed, for this German attack was the result of stealth planning and execution initiated by von Rintelen before he sailed into capture by the British.

At 2.08 a.m. on 30 July 1916, it seemed to the people of New York City that the Battle of the Somme had landed on top of them in an attack so violent and vast that it would stand as the largest assault on American soil for nearly a century. Artillery shells burst over the Hudson River and bullets flew, reports later said, as if pumped out by a thousand machine guns when the *Johnson 17*, a barge carrying 100,000 tons of TNT and 25,000 detonators, exploded in New York Harbor with the force of an earthquake at 5.5 on the Richter scale.

At 2.40 there was a second explosion, and as the one million pounds of ammunition now burst in the conflagration, with bullets landing a mile from where they exploded, New York City and those cities of New Jersey directly across the Hudson River thought they were under attack. Windows crashed from skyscrapers in Manhattan – including every window in J. P. Morgan's headquarters – and were also blown out of the New York Public Library on Fifth Avenue and from every shop along the chic shopping stretch of that avenue. Hotel patrons made their barefooted escape across shards of glass, or through the pools of water unleashed by a broken water main near Times Square.

Ten fire trucks, all racing toward Armageddon, found themselves stuck at Fifth Avenue and 42nd Street, site of a massive traffic jam, while drivers motoring home from a Saturday night in the city across the Brooklyn Bridge swayed in terror with the shock waves that rippled the bridge. Passengers in the Hudson Tubes connecting Manhattan to New Jersey beneath the river feared they were about to be drowned, while prisoners in a jail in Hackensack thought it was being dynamited open. Phone lines between New York and New Jersey went dead, cemeteries saw gravestones and monuments tumble, and people in Philadelphia, nearly 100 miles to the south-west, thought their city had been hit by an earthquake.

Tugboat captains bravely towed blazing barges out to sea. 'Spouting geyserlike pyrotechnics, they drifted across the bay and down upon Ellis Island, sending terror to the hearts of immigrants there.' At 3 a.m., 500 immigrants, many of them refugees from the war for which the ordnance now exploding around them was destined, were evacuated to Manhattan by ferry, their faces lit by the flames as they looked back at the main building on Ellis Island, itself nearly destroyed by the explosion.

Alarm bells rang across Manhattan, and the police, thinking looters and thieves had been unleashed, soon found that it was the shock of the worst blast in the history of the United States that had tripped them. Blowing their whistles and commandeering taxi cabs, they headed downtown, towards the orange glow rising across the Hudson River. One flame that was no more, however, was that in the torch of the Statue of Liberty, just 2,000 feet from the explosion, extinguished to this day when shrapnel tore through Lady Liberty's arm.

When dawn broke and the fire subsided, people flocked to church, praying for the thousands of souls who had surely been lost in the massive explosion. 'When I held mass at six o'clock the church was crowded to the doors,' said Father A. J. Grogan, pastor of Our Lady of the Rosary in Jersey City, 'and I can assure you there were many praying on their knees who had not been inside a place of worship in a long time.' Ammunition would continue to pop and crack, like snipers shooting the wounded, for another three hours after Grogan began his service.

Black Tom Island where the exploding barge was moored and munitions were stored, was really a peninsula, but had been an island once, allegedly

*Workers sort munitions at the Black Tom depot in New York City's harbour*

named after a dark-skinned man once resident there, or because from the air it looked like a black tomcat with its back arched. In 1880, it was connected to Jersey City by a mile-long causeway, to facilitate its function as the terminus for the Lehigh Valley Railroad. When the First World War broke out and the United States started shipping materiel to the Allied cause, the major destination for all goods and munitions was the Black Tom depot, which consisted of 24 warehouses and seven piers. Black Tom housed an estimated one thousand tons of dynamite, TNT, shrapnel shells, nitrocellulose, gasoline and picric acid which would be loaded on to barges, to be hauled out to ships bound for the war.

Miraculously, given the number of men who worked in the warehouses and on the barges of the principal munitions depot in the American northeast, as well as the hundreds of those who squatted on Black Tom, and given the titanic force of the blast, only five people were officially listed as killed, one of them Arthur Tosson, a ten-week-old infant who lived three miles from Black Tom and was thrown from his crib, later dying of shock. Another was Jersey City Patrolman James Daugherty, a 28-year-old cop who arrived on the scene with his partner just after the second explosion. Shrapnel punctured Daugherty's jugular vein, and he was dead by the time he reached hospital. Cornelius Leyden, chief of the private railroad police force, was killed, as was the captain of the *Johnson 17* barge. The body of an unidentified man was also washed up; he was thought to be a nightwatchman.

On 4 August, the *New York Times* reported that a *New York Sun* reporter, 25-year-old Lloyd Wilson, had died of a chill he caught while covering the story on Black Tom Island, and on the 9th, the paper's obituary section announced the death of 27-year-old Mary McGovern, a Jersey City public school teacher, who succumbed to injuries from the explosion. Nearly 100 people were injured, and many more may have died, given the number of undocumented people who were known to have lived on the island.

The physical damage to Black Tom, in terms of lost buildings, infrastructure and goods, was estimated to be $20 million, or more than $450 million today. The first impulse was to call the incident a disaster, and blame the owners and operators of the depot. On Sunday evening, less than 24 hours after the explosions, Jersey City police arrested the superintendent of the National Dock and Storage Company, owners of the pier on Black Tom, as well as the agent for the Lehigh Valley Railroad. The owner of the barge company turned himself in, and the next day arrest warrants went out for the bosses of the Lehigh Valley and Central railroads. All of them were arrested for manslaughter, the initial perception being one of negligence rather than sabotage.

It took the media to focus attention on the possibility of enemy sabotage. On 31 July, the *New York Times* ran a piece reminding readers that there had been 42 explosions connected to munitions in the United States since the war began, and listed them all, as well as those that had occurred in Canada and Europe. On 10 August the paper reported that two Norwegians with pro-German sympathies and drawings of a 'fast submarine' had been arrested in connection with Black Tom. The path of enquiry among those looking into the explosion had shifted, according to the *Times*, with investigators now chasing down information suggesting that the attack on Black Tom 'was the work of alien plotters, acting in this country in the interest of a foreign government'.

On 7 August there had been another explosion in a Lehigh Valley Railroad yard, and this time two men had been seen leaving the scene shortly beforehand, both of them well dressed and one of them wearing white spats. An even more compelling theory for Black Tom being the work of foreign saboteurs was a letter that had been intercepted by the British on its way from the US to Germany. Its author referred to explosions on board ships in Seattle in the spring of 1915, in which he told his parents he had a hand. 'I only wish I could get another chance like that,' wrote 'Otto'. 'I would not hesitate to risk it.'

More than a year before Black Tom, on 15 May 1915, a barge laden with dynamite bound for Vladivostok and the Russian army fighting the Central Powers on the Eastern Front had exploded in Seattle harbour. Franz von Papen had been in the city shortly before that explosion, and had paid the German consul there $1,800.

Despite the fact that von Papen, von Rintelen and Karl Boy-Ed had long been back on the other side of the Atlantic when Black Tom exploded in July 1916, their fingerprints were all over the incident. Von Rintelen had scouted the location himself, part of his expeditions to check out the docks, where he would measure distances and determine mooring spots and escape routes for the saboteurs' motorboats, with inquisitive dockyard guards proving remarkably receptive to paper persuasion: 'wherever a night watchman passed by, or took the liberty of objecting, a few dollar bills gently slipped into his hand ... rendered him as silent as the grave'.

When von Papen had been expelled, he had left his assistant, Wolf von Igel, in charge of dispensing funds for sabotage, but von Igel had been arrested in April in connection with the plot to bomb the Welland Canal. After this, the German agents in Baltimore, Frederick Hinsch and Paul Hilken, became critical to mobilising Germany's war from within.

Hinsch had recruited the slow-witted Slovak immigrant Michael Kristoff while scouting for malleable talent in New York's Penn Station in January 1916. The impoverished and hungry Kristoff, skeletal at six foot three and 147 pounds, readily fell in with the large and intimidating Hinsch, who made him an astonishing offer: he would pay Kristoff $20 a week to watch his bags while Hinsch travelled round the country on business. At the end of the trip, he would get Kristoff a factory job. Hinsch's business was blowing up munitions plants and chemical factories, and upon their return to New York, the factory job he obtained for Kristoff was at the Eagle Iron Works, on the stretch of road leading to Black Tom Island.

Kurt Jahnke and Lothar Witzke, two of the deadliest saboteurs in US history, had come to the enterprise via sabotage on the west coast. Jahnke, 'pimply faced with blonde hair and small weasel eyes', had emigrated from his native Germany to the United States in 1899 as a 17-year-old, and served with the US Marines in their scorched-earth war in the Philippines. By the time the First World War began, he was a naturalised US citizen, and he became one of the principal agents of Germany's consul in San Francisco, Franz von Bopp.

Using his job at the Morse Patrol and Detective Agency as cover, Jahnke provided intelligence on the west coast. Coming under suspicion himself for a massive explosion in November 1915 at a San Francisco munitions factory, he walked into the office of the Secret Service in San Francisco in February 1916 and announced in his German-accented English that he knew about a plot to blow up Mare Island Navy Yard north of the city. The astonished Secret Service agents didn't bite at this cunning offer to become a US operative, and Jahnke left. The following year, Mare Island blew up, and Jahnke was gone.

Lothar Witzke was Jahnke's protégé. Born in 1895, he had been a lieutenant serving on the speedy German cruiser *Dresden* when she was sunk by the British off the coast of Chile in March 1915. Despite having a broken leg, Witzke swam to shore, and was interned in Valparaiso until he escaped early in 1916 and made his way to report for duty to von Bopp in San Francisco. With his naval pedigree, the blonde and dashing Witzke, who took advantage of San Francisco's wine and women, observed ship movements and cargoes while also studying bomb-making with a chemist across the bay in Berkeley.

On the evening of 28 July 1916, the night before the attack on Black Tom, a group of German sailors and spies convened at their regular meeting point, the home of the opera singer Martha Held at 123 West 15th Street, Manhattan. Kurt Jahnke and Lothar Witzke were there too.

A young woman in attendance, Mena Reiss, who was a model for Eastman Kodak and friendly with von Papen and Boy-Ed, later recalled talk of the 'Jersey Terminal', and explosions. She also saw photos and maps of the eventual target. Instead of reporting her fears to the authorities, she took a train to spend the weekend on the New Jersey shore with a friend, where she and her hostess would be jolted awake by the thunderous blast, the plans for which had been finalised in her presence.

Work had stopped at 5 p.m. on Saturday 29 July on Black Tom Island, and, astonishingly, there was no US National Guard or US Marine battalion to protect this major munitions depot, a function again of American neglect and naiveté when it came to realising the country, like it or not, was already at war by virtue of being the main financier and materiel supplier for the Allies. Instead of an armed force of soldiers, only five security guards watched over Black Tom's rail cars and barges loaded with munitions. Two of the guards, Barton Scott and Jesse Burns, worked for the Dougherty Detective Agency, paid for by British taxpayers to guard the docks. Scott and Burns were double agents, also working for Germany's Bureau of Investigation head Paul Koenig, their names later discovered in the coded black book found after his arrest by the NYPD, helped by Room 40's intelligence.

Michael Kristoff left his aunt's house in Bayonne, New Jersey, at 11 p.m., telling her that he was going to pick up his pay from work, something she thought odd given the hour. In fact Kristoff was going to walk through the gate to the Black Tom pier and plant an explosive on a rail car.

Jahnke and Witzke were in a dinghy, having launched from the side of a German merchant ship, rowing toward the barges anchored at Black Tom. By half past midnight, the trio had planted their explosives and made their escape. Fifteen minutes later, a barge captain returning from a night out was chatting with Jesse Burns when the bribed detective noticed flames in a rail car. The greatest act of sabotage in US history had begun its fatal trajectory.

On 4 August 1916, Paul Hilken threw a celebration for the Black Tom plotters at Hotel Astor, just off Times Square. Hilken, who funnelled money to German agents, and Hinsch, who planned their attacks, enjoyed the summer air of the restaurant's roof-top garden. Hilken tried to pry details out of Hinsch as to who planted the bombs, and how, but Hinsch, revealing just how difficult it would be to catch any dedicated network that maintained radio silence, told Hilken that it was 'better that you don't know too much'. Hilken's agreement lay in the two $1,000 dollar bills he handed to Hinsch as his take for a job well done. It would take the United States more than two decades to make the case for just who and what was behind the Black Tom explosion, which they did with the help of Blinker Hall's private papers from Room 40, which he unsealed especially for the American prosecutors in 1925. Room 40 had not intercepted anything explicity stating the sabotage plan at the time, but intercepts of messages sent from German agents in North America to Berlin in 1918 established that German operatives were involved in planning and staging the attack. In the immediate aftermath of the explosion, President Woodrow Wilson's mind was on his re-election campaign, the slogan for which was 'He Kept Us Out of War'and despite the calls for war, the US response was to blame the transporters of the goods, and then, in September, to convene a public hearing at the direction of the Secretary of War Newton Baker, to reposition munitions barges in New York Harbor, in new locations, and 'the distance such explosive carrying barges should be kept apart'.

Given the attacks on the United States from within and at sea, the question of just how long Wilson's neutral stance could remain was bigger and more pressing to the Allies than ever. While Blinker Hall and his team had certainly worked to bring the USA to join the Allied cause, they had not suppressed foreknowledge of Black Tom to speed up the Americans' entry into the war. Yet, if a massive attack on America's nascent 'capital city of the world' wouldn't spur President Wilson to action, just what would it take to bring America to the fight, now that the conflict had been so fatally brought to America?

# Chapter 14

# THE ZIMMERMANN TELEGRAM

Very early on the morning of 17 January 1917, hope for the Allied cause appeared in the form of a telegram handed over to Nigel de Grey and Dilly Knox, who were manning the night watch in Room 40. They quickly realised that the message was encoded in the German diplomatic code 0075, which they had begun intercepting between Berlin and the US embassy in November 1916.

The duo worked on the telegram for hours, identifying the message's recurring groups and then cobbling together the beginnings of a decode. De Grey, whose German was better than Knox's, realised that this note destined for Mexico – to be sent via Count Johann von Bernstorff, the German ambassador to the USA – from German Foreign Secretary Arthur Zimmermann, was Room 40's version of the Holy Grail.

Though the message was only partially decoded, the slender, aristocratic de Grey ran all the way down the corridor and into the office of his boss, Blinker Hall.

'Do you want to bring America into the war, sir?' he asked.

It was a question of pure rhetoric, for everyone in Room 40 knew that American military muscle was the Allies' greatest, and perhaps last, hope of victory. When Hall acknowledged the obvious, de Grey produced his triumph: 'I've got the telegram that will bring them in if you give it to them.'

The partially decoded telegram had revealed enough for de Grey to realise that the plan detailed within it would send America into bellicose fury. And even though it came from Germany, the plan to keep America out of the war in Europe involved the country that made the USA most nervous, the country right on its own border: Mexico.

Arthur Zimmermann knew about American paranoia toward Mexico, too. When he was appointed Germany's Secretary of Foreign Affairs on 24 November 1916, the 52-year-old diplomat became the first non-aristocrat to hold that exalted post, getting there through his wits and cunning. Yet the amiable, forthright Zimmermann, with his bushy reddish-blonde moustache and his duelling scar, was hardly a peasant with a pitchfork who suddenly found himself as Germany's steward of foreign affairs. Like the Junkers class who governed Germany, he too was from Prussia. After earning a doctorate in law and practising briefly, he joined Germany's foreign service in 1893, when he was 29 years old. He was consul in China during the Boxer Rebellion, and wound up back in Berlin as Under Secretary of State in 1911. In 1914, as acting foreign secretary, he agreed with Kaiser Wilhelm and Chancellor Bethmann-Hollweg that Germany must

LIBRARY OF CONGRESS, WASHINGTON

*German Foreign Secretary Arthur Zimmermann*

ally with Austro-Hungary after the assassination of the Archduke Franz Ferdinand, and drafted the telegram announcing that intention. In 1916, he worked with Roger Casement to foment revolution in Ireland.

When his predecessor as foreign secretary refused to support a renewal of unrestricted submarine warfare, which had been suspended in late 1915, Zimmermann, a whole-hearted supporter of Germany's military leaders, got his job. On 31 January 1917, US Ambassador James Gerard was summoned to his office to be officially informed of Germany's decision. Zimmermann told Gerard that this strategy 'was a necessity for Germany, and that Germany could not hold out a year on the question of food. He further said, 'Give us only two months of this kind of warfare and we shall end the war and make peace within three months.'

Germany's food situation was central to the desperate military decisions the country made during the winter of 1917–18. Shortly after the Germans resumed their U-boat free-for-all on the Atlantic, and the US responded by breaking off diplomatic relations, a member of the Prussian Diet rose in alarm to report to the assembly that 'the mortality among elderly people is increasing at a terrible rate'. He also declared that epidemics were spreading among the weakened population, that suicides were increasing and that 'parents are killing their children rather than see them suffer the pangs of unsatisfied hunger'. If the war lasted another year, Germany itself would die of famine.

Zimmermann's incendiary telegram was then, in terms of war strategy, potentially brilliant – after all, the Germans had been using the US as their far western front since the beginning of the war. Rather than being a direct attack, Zimmermann's note was to be delivered to the Mexican president only if the United States entered the war due to Germany's unleashing of the U-boats on all Atlantic shipping. It called for an alliance between Mexico and Japan, heavily subsidised by Germany, to wage war on the USA, with the victory prize to be territories the United States had won from Mexico. Zimmermann could not have picked a country more likely to bring the US into the war.

*

Mexico had always been a problem for the United States, and vice versa, with territorial wars, skirmishes and horse trading resulting in the USA winning Texas as a state in 1845, and New Mexico and Arizona in 1912. Two years later, President Woodrow Wilson had sent American sailors and marines into Veracruz on a six-month campaign to prevent the German government from sending arms to the Mexican president, Victoriano Huerta, with the resulting violence killing 19 Americans and 129 Mexicans.

In 1915, Mexican raids had killed 21 Americans as part of the 'San Diego Plan', a manifesto drawn up in the small Texas town of San Diego by President Venustiano Carranza to create an extraordinary liberation army of Mexicans, African-Americans, aboriginals and Japanese. Under the red and white banner of 'Equality and Independence', this rainbow coalition would slaughter every Anglo male over the age of 16 in their quest to reclaim Texas, Arizona, Colorado, New Mexico and California to create an independent republic. The raids – and the revenge plan – would be called off once the US government recognised Carranza as the legitimate president of Mexico, which it did in the summer of 1915. And that got General Francisco 'Pancho' Villa angry.

Francisco Villa had been born José Doroteo Arango Arámbula, the son of poor peasants in the state of Durango. After his father died, Villa became a sharecropper, then abandoned that for the riches of banditry. He soon became part of a 'super group' of *bandidos*, and ironically, avoided execution upon capture because of the intervention of a powerful landlord to whom Villa had sold stolen goods.

His punishment was to serve in the federal army, but in 1903, he killed an officer and headed to the state of Chihuahua using the name of his paternal grandfather. The man 'Francisco Villa' had been born, and for the better part of the next decade would re-invent himself as a kind of Mexican 'Robin Hood', leading his *bandidos* on raids against the bad hacienda owners in the name of the oppressed people.

When the Mexican Revolution began in 1910, Villa was in the thick of it, fighting the dictator Porfirio Diaz with the pro-democracy forces, and winding up alongside General Huerta. Huerta, however, was jealous of

Villa, and after accusing him of horse theft, insubordination and outright insurrection, Villa found himself standing in front of a firing squad. He was reprieved just in time – so the romantic story goes – by a telegram from Madero commuting Villa's sentence to prison, from which Villa escaped in time to join the loathed Carranza in a war against the even more despised Huerta – now dictator of Mexico after the murder of Madero.

Villa fought with Carranza to depose Huerta, along the way getting himself elected governor of the state of Chihuahua in 1913. As governor, Villa printed his own currency, and such was his stature that it was accepted at par at banks in Texas. Villa ordered his paper money also to be taken at par with Mexico's gold pesos, and in a move that would further cement his 'Robin Hood of Mexico' reputation, forced wealthy landowners to give loans – and land – so he could pay and feed his troops, and compensate their widows and children. He also took gold from banks, and true to his bandit origins, took wealthy hostages if the banks were less than forthcoming about where they kept the gold.

The gregarious, lavishly moustachioed Villa, a teetotaller who had honed his peasant intelligence by learning to read and write in prison, was as quick to laugh as to pull the trigger – once shooting dead one of his soldiers for being drunk and loud while Villa was giving a journalist an interview. Indeed, his larger-than-life persona attracted a Hollywood film crew who followed him around to document his exploits – though Hollywood wasn't there to witness Villa's atrocities that finally provoked the USA into action. On 11 January 1916, Pancho Villa's men hauled 17 American mining engineers from a train in San Ysabel, lined them up and shot them (one man faked death and escaped), prompting the US to put El Paso, Texas, under martial law to prevent its enraged citizens from crossing the border to take revenge.

Two months later, while darkness still cloaked the dusty border town of Columbus, New Mexico, Pancho Villa launched a raid with 450 of his mounted soldiers, known as 'Villistas'. For more than an hour the Villistas, hollering *'Viva Villa! Muerta a los gringos!'*, wreaked deadly havoc, setting fire to homes and businesses in the town of 700, and shooting people where they found them. Eighteen Americans and 80 Villistas would die in the attack, with four captured Villistas later hanged.

President Woodrow Wilson – who had successfully managed to avoid committing the US to a disastrous war in Europe – now did the very thing the Germans had always hoped he would do: get sucked in to Mexico and expend America's military energies there. Wilson knew he had to take military action, especially in an election year, with high-profile people such as former president and pro-war agitator Teddy Roosevelt mocking his benighted pacifism in person and in print. Wilson's response to Villa's latest outrage would send American troops deep into Mexican territory under the command of the man who would eventually lead the US army in France. And it would give Germany one last desperate idea to keep America out of the war once and for all.

With the war in Europe generating death on an industrial scale but no clear victor, and given the enthusiasm of Germany for using the USA as a 'third front', it was no surprise that Villa's attacks on the US were helped by German money. At least $340,000 of German money was funnelled by German agent Felix A. Sommerfeld from a bank account in St Louis to fund arms for Villa.

The war against Pancho Villa launched by the US in April 1916, known as the 'Punitive Expedition', sent more than 14,000 American troops 450 miles deep into Mexico in pursuit of a man who had recently been their trusted ally and receiver of US armaments. Indeed, his military tactics had been so admired by the US that not only did the army study them, but Villa had, in happier days, been invited to the army command centre at Fort Bliss, in El Paso, to meet John J. Pershing, the man who, unbeknownst to them both, would soon be hunting him down.

Pershing, tasked with the mission to disrupt and end Villa's campaign, and either capture or, better still, kill him, was a 55-year-old career soldier with an iron jaw, ramrod posture and sharp, unsentimental eyes. He was nicknamed 'Black Jack' while teaching at West Point, due to his service with African-American Buffalo Soldiers first in the Indian Wars, and then again in the Spanish-American War – a nickname that had been softened

*General John Pershing in France*

from something far more offensive. Pershing's mission in Mexico was highly sensitive. The USA had the tacit support of Mexican president Venustiano Carranza, but if Pershing's forces pushed too hard, a total war could easily result, and the US wanted to punish Villa, not ignite the bone-dry tinder that was Mexico.

Despite Carranza's promise to let Pershing's forces use the Mexican Northwestern Railway, the Mexican army blocked the free flow of troops and supplies. They also repeatedly cut the US army's telegraph wires. Given the hostility of the environment both natural and political, Pershing put a premium on intelligence, and it was here that the US army finally joined the intelligence war.

Pershing organised his own field intelligence network, and started an information department, whose agents – Mexican, Japanese, and 20 Apache scouts – worked to track Villa and infiltrate his organisation of *bandidos*, who would routinely divest themselves of their guns and blend in with the local population, sometimes even watching movies with unwitting American officers in the cinemas.

Pershing also used 'radio tractors', trucks equipped with radio sets, to listen in on Mexican communication. Both the government and Villa's forces transmitted by wireless, and American intercepts of coded Mexican messages would be sent to Captain Parker Hitt, who wrote the US army's first book on cryptology; his *Manual for the Solution of Military Ciphers* was published in 1916 at Fort Leavenworth. Hitt used the Mexican Army Cipher Disk to decode the messages, a method involving four numerical alphabets placed on a revolving disc. 'By tapping the various telegraph and telephone wires and picking up wireless messages,' Pershing wrote in his report of the mission, 'we were able to get practically all the information passing between various leaders in Mexico.'

It wasn't enough, and the campaign was all over by February 1917. The last American cavalry mission had failed to capture its prize, but it had learned much about conducting a new kind of war on foreign soil. The intelligence that Pershing had employed would become critical in his next campaign, in Europe. The Yanks – to the great relief of the Allies – were about to head 'over there'.

One of the great 'what ifs?' of history lies in the German plan to deliver the Zimmermann telegram not via cable, tapped into and intercepted by the British, but by submarine. On her second trip across the Atlantic to a US port (the first, in 1916, had been to Baltimore), the German cargo submarine *Deutschland* docked at New London, Connecticut, on 2 November 1916, bringing 750 tons of dye stuffs, chemicals, and medications against polio, along with Germany's 0075 code book in a sealed diplomatic pouch that was delivered to the German embassy in Washington. It was this code book that would allow Count von Bernstorff and his staff to decrypt the Zimmermann telegram.

The *Deutschland* was scheduled to sail again for America on 15 January 1917, carrying Zimmermann's extraordinary offer to Mexico.

*The U-boat* Deustchland *arriving in Baltimore harbour, July 1916*

When Germany resumed unrestricted submarine warfare in February, the *Deutschland* was drafted back into service, and her third mission to America was aborted. Had it not been, it is tantalising to speculate that the course of the entire world war – and the one to follow it – might have been changed. But it was via cable that Zimmermann's telegram came, and while it filled Room 40 with robust hope, this particular communication had to be handled very carefully.

The problem facing the British was twofold: they had to counter the possibility that the Americans would think the telegram a hoax; and they needed to conceal the fact that the British were reading German dispatches to Washington via the US telegraph cable, and so reading American dispatches as well. They needed to present the telegram to the Americans by disguising its source.

All transatlantic cables passed by the east coast of Ireland or the west coast of England, and were relayed to the central telegraph exchange and copied by the censorship office. Germany sent its telegrams to America by two routes, the first of them evidence of the United States' native generosity and total naiveté when it came to sophisticated acts of war. At the end of

1916, Colonel Edward House, Wilson's counsellor and confidant and White House power-broker, had arranged to let the Germans send their telegrams directly to him, via the American cable, in the interest of brokering peace with von Bernstorff in Washington.

The second route was known as the Swedish Roundabout, a method devised in 1915 when Britain complained to neutral Sweden that while, yes, it was reading Sweden's telegrams, the Swedes were violating the laws of war by sending German messages through their cables to Washington. The Swedes admitted guilt, and then just redirected Germany's messages to Buenos Aires, handing them over there to the Germans for transmission to Washington. This time, Room 40 did not complain. They just kept reading.

Blinker Hall knew that with no small irony, the answer to his problem lay in Mexico. Zimmermann had sent his telegram to von Bernstorff via both Washington and the Swedish Roundabout. It was von Bernstorff's duty to transmit the note to Heinrich von Eckardt, Germany's minister in Mexico. If Hall could somehow get his hands on the copy of the telegram in Mexico, the version sent from von Bernstorff to Mexico would have a different time stamp and serial number than the one intercepted by Room 40, and so it would appear to the Americans – and to the Germans – as if whoever had discovered the telegram had only done so in Mexico and so Room 40's hands would be 'clean'. And best of all, the German embassy in Mexico didn't use the 0075 code. Von Bernstorff would have to recode the message in a code that Room 40 knew well. This would allow them to solve the telegram with certainty.

It was a brilliant idea, but how to pull it off? There is a wonderful romantic story about British agent Thomas Hohler securing the freedom of a British printer facing imminent execution by the Mexicans under suspicion (wrongly) of forging banknotes. In gratitude, the printer had his brother steal the telegram from the Mexican telegraph office, where, *mirabile dictu*, he just happened to work.

In all likelihood, it was good old-fashioned graft or threat of blackmail that got the telegram into Hohler's hands, and then into those of Room 40, who finished decoding it. Now all that remained was to convince the Americans that this telegram meant war. So Blinker Hall summoned Edward Bell to see him.

Hall liked and trusted Eddie Bell, the Second Secretary of the US embassy in London. He had used him as unofficial liaison for intelligence matters that he wanted to bring to the attention of American ambassador Walter Hines Page. Page, who had been editor of the *Atlantic Monthly*, as well as a partner at Doubleday, Page & Company, publishers, was a devout Anglophile who believed that Britain was fighting for democracy. Hall exploited this by giving Bell details of his interrogation of Franz von Rintelen, as well as von Rintelen's papers, and those confiscated from the courier James Archibald, which included a plan from the Austrian ambassador to the US, Constantin Dumba, to disrupt the American steel and munitions industries with strikes.

On 19 February 1917, a grey Monday in London, warmer than it had been after an unusually cold winter, Ambassador Page despaired of America and her pacifist president. 'I am now ready to record my conviction that we shall not get into the war,' he confessed to his diary. '[Wilson] is constitutionally unable to come to the point of action.'

Over at the Admiralty, Blinker Hall hoped that he was holding the smoking gun that would shake Wilson out of his peace dream into the reality of the war he and his team were so arduously fighting. Bell's initial reaction was one of incredulity when he saw the decrypted note from Zimmermann promising Mexico large chunks of the American south-west should it wage war on the USA. 'Why not Illinois and New York while they were about it?' he thundered. But then he calmed, and wondered if this document might be a forgery or a hoax.

Hall knew that he had to proceed with caution, as Bell's reaction was a barometer to those who would doubtless follow. He explained that British agents had discovered the telegram in Mexico, and after bringing Ambassador Page into the conversation, the trio concluded, at Page's insistence, that the message would carry maximum weight in Washington if the British government formally presented him with the decrypted telegram. Page was well aware that the British blacklisting of American companies who were accused of doing business with the enemy, as well as the interception of US mail on the North Atlantic and the rough justice meted out to Irish rebels, had made the need to join the Allied cause less clear-cut to many of his countrymen than it was to him.

On Friday 23 February, the patrician and sanguine Foreign Secretary Arthur Balfour was in a state of excitement as he received Ambassador Page at the Foreign Office. He would later say that the moment when he handed over the document that would change the face of the war was 'the most dramatic of my life'.

Page stayed up until 3 a.m. drafting a careful cover letter to Woodrow Wilson to send with the decrypted telegram, which he transmitted later that day. In his letter, the masterwork of Hall is again evident, for though Room 40 had been in possession of the telegram for more than a month, Hall had clearly told Page otherwise, resulting in the ambassador informing his president that the British had 'lost no time in communicating it to me to transmit to you ... in view of the threatened invasion of our territory'.

Though Zimmermann's telegram threatened a Mexican invasion of the US only if the US joined the Allied cause, the point was moot when Secretary of State Robert Lansing added insult to injury by revealing to Woodrow Wilson that the Germans had sent their message of war via the American cable that the US had so generously allowed them to use in the name of peacemaking. 'The President two or three times exclaimed "Good Lord!" ... and showed much resentment at the German government for having imposed upon our kindness in this way and for having made us the innocent agents to advance a conspiracy against this country.'

Despite his naiveté about the dark arts of intelligence, Wilson also wondered if the document might be a fake. State Department counsel Frank Polk had already thought of that, and managed to arm-twist Western Union to release the copy of von Bernstorff's transmission to the German mission in Mexico City.

On 1 March, newspapers in the United States and around the world trumpeted the treachery: 'GERMANY SEEKS ALLIANCE AGAINST US; ASKS JAPAN AND MEXICO TO JOIN HER; FULL TEXT OF HER PROPOSAL MADE PUBLIC;

WASHINGTON EXPOSES PLOT' shouted the *New York Times*' front page. The White House had leaked the Zimmermann telegram to the Associated Press, and while the *New York Times* felt compelled to say – again with abounding irony – that it hadn't authenticated the contents, it had no hesitation in revealing the depth of Germany perfidy. It seemed to outraged Americans that it was their own government who had broken open the ugly truth of the German betrayal, which was exactly how Blinker Hall wanted it to appear.

German-Americans immediately saw the whole affair as a provocative fiction. The *New Yorker Staats-Zeitung* editorialised: 'The passions of the American public that still doesn't want to hear of war must be aroused so that it may attain that condition forced by similar means on the people of Italy, Great Britain, and Rumania.'

The *Fatherland*'s proprietor and inveterate anti-Allied provocateur George Sylvester Viereck was more bluntly incredulous, calling the telegram 'obviously faked' for the simple reason that 'it is impossible to believe that the German Foreign Secretary would place his name under such a preposterous document'.

On 3 March, Zimmermann put any doubt to rest, astonishingly confessing to plotting the creation of a Japanese–Mexican invasion of the USA. 'When I thought of this alliance with Mexico and Japan I allowed myself to be guided by the consideration that our brave troops already have to fight against a superior force of enemies, and my duty is, as far as possible, to keep further enemies away from them,' he declared in a speech. 'That Mexico and Japan suited that purpose even Herr Haase [Hugo Haase, a German socialist politician and pacifist] will not deny. Thus, I considered it a patriotic duty to release those instructions, and I hold to the standpoint that I acted rightly.'

Even with Zimmermann's admission, the United States – well aware that it was on the march to war – wanted to silence the doubters that still remained. So in one of the most baroque – and, were it not for the outcome, comic – reversals in the history of codebreaking, the US State Department cabled Ambassador Page to ask if the British would permit someone from the US embassy 'to personally decode the original message which we secured from the [Western Union] telegraph office in Washington ... and make it possible for the department to state that it had secured the Zimmermann note from our own people'.

Blinker Hall was only too happy to go along with the ruse: there was no one in the US embassy who had the slightest idea how to decode the Zimmermann telegram. But in the spirit of the elaborate theatre needed to convince the American people that this was not a slick foreign plot, Eddie Bell came to the Admiralty and looked on as Nigel de Grey decrypted the telegram for him, having no choice but to believe that the jumble of numbers on the telegram meant what de Grey said they did. Ambassador Walter Page accepted at face value Blinker Hall's claim that it would be pointless to burden the US with the German code book because it 'would be of no use to us as it was never used straight but with a great number of variations which are known to one or two experts here. They cannot be spared.'

With this last bit of theatrics, Hall had primed the US for war without giving up the fact that the British were reading secret American communications. Walter Page, who wanted to live another 20 years to be present for the unveiling of Hall's secrets, wrote in a letter to President Wilson: 'the man is genius – a clear case of genius'.

By the middle of March, alarmist reports were coming out of Mexico that the country was swarming with German troops, which it was not, though German spies had found their way south to avoid being on American soil when war was inevitably declared. On the 18th, three American merchant ships, *City of Memphis*, *Illinois* and *Vigilancia*, were sunk by U-boats, with 38 men missing. The *New York Times* also reported that Germany had announced it had sunk nearly 800,000 tons of Allied shipping since it resumed unrestricted warfare in February – a total of 292 hostile ships and 770 neutrals.

In early April the galvanised United States Congress authorised Woodrow Wilson to go to war. In a 36-minute speech, Wilson laid out what the country had now accepted: a state of war existed between the United States and Germany. 'The world,' he famously said, 'must be made safe for democracy,' but in the end, what propelled the United States, finally, into the First World War was the brilliant decryption and dissemination of a document that threatened its own territory, and its own democracy. The United States was fighting a war in Europe so that it wouldn't have to fight one in Mexico. Pershing would take what he had learned from his Punitive Expedition, especially in the field of intelligence and codebreaking, and use it in the war in France. With a particularly American twist.

# PART III

PART III

# Chapter 15

# MASON, MULES AND ROUNDABOUTS

You would have thought that the disastrous fallout from the Zimmermann telegram would have deterred the Germans from meddling in Mexico: on the contrary, now that America was officially their enemy, there was even more reason to use it as a base for action against the USA.

Initially, the Germans continued to try and nudge Mexico into open conflict, offering weapons, logistical support, money and debt relief. Kurt Jahnke, who had been instrumental in the bombings carried out on American soil – most notably Black Tom – crossed the border to take charge of subversion and sabotage with 'an available credit of 100,000 Marks per month'. Instructions relayed to him by telegram from Berlin that passed via Madrid were intercepted and decoded by Room 40. Of particular interest were plans to use German submarines to target trade routes: 'undertakings against the Panama Canal are highly desirable. If a good opportunity presents itself, the corn ships sailing from Australia to America should be attacked.'

To facilitate such endeavours, Jahnke was ordered to find places that could serve as secret refuelling sites for U-boats. On 9 June 1917 Room 40 picked up a message sent to the German legation in Mexico City requesting Jahnke 'to prepare as rapidly as possible a point ... for submarines on the Mexican coast'.

Though there was some considerable panic in America at the prospect of U-boats raiding the West Coast, the threat never materialised. Then, during August 1917, the emphasis of German policy changed: while Mexico would still offer refuge to agents fleeing American justice, such as Anton Dilger, the biochemical weapons expert, economic penetration became the priority. The aim was to deprive the Allies of resources such as lead, copper and oil by either acquiring control of companies through share purchasing or setting up rival corporations.

Though large sums were made available, these measures had little impact. A plan to burn the major oil fields at Tampico, which had been abandoned during the Zimmermann scandal, was resuscitated. A German agent was given the job and told that 'if arson not possible, at least disrupt loading and capacity to supply'. However, von Eckardt, the German resident minister in Mexico City, worried that his hosts were beginning to lose patience with German subterfuge, blocked the operation.

A project that did go ahead was the construction of a wireless station with a receiver/transmitter capable of sending messages directly to Europe. Hall's agents in Spain discovered that the Germans had purchased a quantity of high-powered audion valves, which were able to receive long-distance signals and amplify wireless communications, and had shipped them to Mexico. Equipped with the valves, the wireless station, located at Ixtapalapa, just outside Mexico City, was soon up and running and relaying signals to Madrid that were then rebroadcast to the main German station at Nauen. These developments deeply concerned Hall, who feared that such a powerful station would allow the Germans to coordinate their subterfuge in both North and South America without resorting to the telegraph, and thereby cut Room 40 out of the equation.

Hall decided to send his favourite field agent, the author and adventurer A. E. W. Mason, who had distinguished himself in Spain with his resourcefulness, bravado and cunning, to Mexico City: his priority, to sabotage the wireless station. Armed with a courier's passport that allowed him to travel 'freely without ... hindrance', Mason left Liverpool on 19 October 1917, passed through Washington where the passport was countersigned by the British ambassador, and arrived in Mexico City in November.

Posing as a lepidopterist (butterfly collector), and carrying suitable nets and equipment to maintain his cover, Mason soon found the Ixtapalapa site. After making discreet enquiries, he discovered that German wireless officers from ships interned at Veracruz were coming ashore every night at 11 p.m. to man the wireless station until the following morning.

Having returned briefly to London to get the green light from Hall, which he duly received, Mason arrived back in Mexico and set about buying up all 11 audion lamps still in circulation there, before turning his attention to Ixtapalapa. His first move was to establish how many lamps were at the station. To assist him he recruited two high-ranking police officers and a burglar, 'three Mexicans of worth', and befriended a local dignitary who agreed to join the conspiracy. According to Mason, whose operational notebooks provide the main evidence of his clandestine work in Mexico, this gentleman of 'high position' invited the captain of the 40 soldiers who guarded the facility to dinner; also in attendance were two of Mason's team. A convivial, boozy evening ended in an invitation to visit the wireless station.

The next morning Mason's team reported back: 'the soldiers were in a large room on the ground floor. The receiving apparatus was upon the first floor, and there were 13 audion lamps in use', plus three spares. They also revealed that there was a window at the end of the large receiving room that afforded an easy drop into a garden enclosed by a fairly high wall. This, however, 'would present no particular difficulty to the expert amongst the party', i.e., the burglar!

After Mason had his agents make another visit to the station, where they were allowed to take photographs, he became convinced his scheme was feasible. The captain of the guards was to be invited to another dinner. Mason's men would strike in his absence, before the German wireless operators arrived for their shift.

Mason's helpers arrived at the station bearing gifts: several jars of *pulque*, the Mexican equivalent of beer. Having got some of the soldiers thoroughly drunk, a fight was instigated over an alleged insult, which escalated into a general brawl 'with sticks and fisticuffs'. Taking advantage of the confusion, Mason's men 'darted up the stairs into the receiving room, twisted off four of the audion lamps, smashed the rest, jumped out of the window, climbed the wall at the appropriate spot, and dropped into the motor-car which was waiting in the road just beneath. The car was then driven back to Mexico City as rapidly as the abominable roads would allow.'

This left one audion lamp unaccounted for. Determined to complete the job, Mason got himself invited to yet another party at the station. During the spirited revelries, he pretended to get drunk. Feigning sickness, he slipped out, tracked down the lamp and hid it in his jacket, then made his getaway in a chauffeur-driven car.

For the rest of 1918, Mason produced a newspaper, *El Progreso*, filled with disinformation and Allied propaganda, and focused on the movement of suspected German agents back and forth across the border; he ran down one of the Black Tom bombers, and had him arrested in the States, where he was court-martialled. What else he got up to while he was in Mexico is impossible to say for sure. He did not recycle any of his escapades in his novels, as he did with his Spanish adventures. However, two anecdotes he told friends after the war have survived, thanks to Mason's biographer.

The first is particularly chilling. Mason claimed to have arranged the assassination of German agents who were trying to set up a portable wireless transmitter in a remote part of Mexico. Mason, who remarked that 'you can get a man killed out there for five shillings', put word out on the street that they were carrying diamonds, with the result that 'no more was heard of them or their wireless'.

The other anecdote concerned an attempt on his life: 'Mason received a message that confidential information of utmost value would be divulged if he would go to a rendezvous at a certain address. He agreed, but ever suspicious of Teutonic guile, he thought it advisable to reconnoitre on his own account; and so, his identity suitably disguised, he visited the place – to find it situated on a remote and ill-favoured by-way with a nail-studded door and barred lattices.

Suspecting that he was being lured into a trap, Mason arrived at the location shortly before the appointed time and hid at a window on the opposite side of the street. Watching from his vantage point, he saw 'a couple of tough-looking individuals' go into the building 'ten minutes before ... the hour at which he himself was expected'. Reflecting on this close brush with death, Mason remarked that 'if I had gone in ... I should never have reappeared'.

Though Mason did not mention these incidents in his official reports, they have the ring of truth about them. Certainly he was involved in a murky and murderous business. The brutal and vicious civil war that was tearing Mexico apart at the time made it an extremely treacherous place to be. Yet it is just as likely that he invented them to entertain and impress his friends. In the world of espionage, fact and fiction often go hand in hand. What cannot be denied is that Mason was one of the most effective agents Hall ever employed.

By exposing the Zimmermann telegram in the way that he did, with a cover story provided by the Americans, Hall overcame one of his greatest fears: that the existence of Room 40 would become public knowledge. It also showed him that he could use Room 40 intelligence more overtly without raising suspicions about its true origin. As long as the Americans agreed to play ball, further damage could be done to Germany's international credibility and its capacity to wage war. The question was where to strike next.

The Swedish Roundabout, which allowed Berlin to communicate with its embassies by using Sweden's diplomatic channels of communication, was an obvious target. By revealing it to the world, Hall would provide another example of German perfidy and hopefully shame the Swedes into abandoning their pro-German government at the forthcoming elections, replacing it with the pro-Allied alternative. Simultaneously, it would undermine Sweden's role in helping Germany evade the naval blockade: large quantities of contraband goods were being exported to Sweden and then re-exported to Germany.

These gains, however, had to be weighed against the fact that exposing the Swedish Roundabout would deprive Room 40 of the valuable information gained from it.

Hall, therefore, sought to maximise the impact of further disclosures, and his thoughts turned towards Argentina, a key source of foodstuffs essential to both the Allies and the Germans. From Argentina came wheat, grain and beef, the demand for which spiralled as the war progressed. In 1913, 6.7 million tonnes of shipping operated out of Buenos Aires; by 1917, that figure had leapt to over 20 million.

Argentine wheat was vital to the Allies' survival. At the beginning of 1917, they attempted to secure its supply: shortages in Britain were becoming acute, rationing was on the way. Any interruption to this trade would be fatal. Desperate, they agreed to loan the Argentine government 200 million pesos in exchange for 2.5 million tonnes of cereal. A similar situation pertained to Argentine beef. In late August 1914, a deal was done to guarantee monthly deliveries of 15,000 tonnes of beef to Britain, 80 per cent for the army, 20 per cent for domestic consumption. By spring 1916, 25,000 tonnes per month were being imported; during 1917, double that amount.

Despite these bilateral agreements, Argentina continued to supply Germany with these commodities too, defying the blockade by shipping them to Holland and Scandinavia, from where they were then transported to the Fatherland. The Germans also had an extensive network of agents in Argentina who helped to maintain this trade and thwart economic cooperation with the Allies. Meanwhile, the return of unrestricted submarine warfare meant that Argentine merchant ships were now a legitimate target. Clearly matters were coming to a head.

For some time Hall had been frustrated by his lack of traction in Argentina: he simply didn't have the resources or manpower available to match the German presence there. What he did have was Room 40. As it happened, the German minister in Buenos Aires, Count Luxborg, used the Swedish Roundabout to communicate with Berlin. As these messages passed via British cables en route, Room 40 was able to intercept and decode them.

From May 1917, the codebreakers began work on this traffic – aided by Malcolm Hay's team at MI1(b), who provided material not only on Argentina but also on Brazil, Chile, Uruguay and Peru. Deciphered messages were forwarded to Robert Lansing, the US Secretary of State. Washington, still blissfully unaware that Room 40 had been reading its communications for several years, responded in kind: all the telegrams sent by Ambassador von Bernstorff to Berlin that had travelled through US diplomatic channels were forwarded to London. Room 40 decoded them as well. Before long, Hall had accumulated a collection of messages that would seriously embarrass the Germans in Argentina.

He outlined his strategy to Walter Page, the US ambassador in London, who had been deeply impressed by his handling of the Zimmermann affair. Hall's aims were clear: to force Argentina and Sweden to abandon Germany; to influence other neutral countries, Spain in particular; to undermine Austrian, Turkish and Bulgarian faith in their ally; and to further depress German morale.

Page, enthusiastic about Hall's plan, sent a secret telegram to President Wilson at the end of August outlining the way forward:

> Admiral Hall … has given me a number of documents comprising German cipher messages between German diplomatic offices and the Berlin Foreign Office, chiefly relating to the Argentine … the British government hope that you will immediately publish these telegrams asking that their origin be kept secret as in the case of the Zimmermann telegram. I have the cipher originals and am sending them to you by a trustworthy messenger who will deliver them into your hands about 12–15 September. These telegrams also prove that Sweden has continuously used her legations and pouches and her code to transmit official information between Berlin and German diplomatic offices.

Wilson agreed to the subterfuge. It was then merely a matter of choosing which messages to release to the press. Count Luxborg's telegrams to Berlin demonstrated the contradictory position he was forced to adopt due to unrestricted submarine warfare. On the one hand, sinking Argentine merchant shipping was necessary if the campaign was going to succeed. On the other, it inevitably worsened relations with the host nation. With stakes that high, it is no wonder that Luxborg's attitude alternated between aggression and caution.

A decoded message sent by Luxborg on 19 May regarding two Argentine steamers, the *Oran* and the *Ginzo*, 'which are now nearing Bordeaux', suggested that either the ships 'be spared if possible' or 'sunk without a trace being left'. Another communication, dated 9 July, was equally muddled: 'As regards Argentine steamers, I recommend either compelling them to turn back, sinking them without leaving any traces, or letting them through.'

When a selection of these messages appeared on the American front pages, there was an immediate outcry. Luxborg's case was not helped by the contemptuous tone he adopted when referring to his Argentine colleagues: they were 'under a thin veneer, Indians', while the acting minister for foreign affairs was described as 'a notorious ass and anglophile'.

The Argentine government promptly severed diplomatic ties with Germany and expelled Luxborg. The reaction in neighbouring Brazil was even more forthright. Up to that point, Brazil had been exporting large quantities of coffee to Germany via Holland and Scandinavia. The increase in this trade compared to pre-war levels was dramatic. Between August and December 1913, Germany had bought 173,000 bags of coffee from Brazil. During the same five-month period in 1915, it received 1,795,000 bags. Though the blockade was beginning to bite, the quantities getting through were still substantial. The publication of Luxborg's messages put a stop to that: Brazil declared war on Germany. Its citizens and troops would have to make do with ersatz coffee, made out of acorns, for the rest of the conflict.

In Sweden, the exposure of the Roundabout had the desired effect. At their elections in early 1918, the pro-German conservative government was beaten by the pro-Allied liberal/socialist Democrat Coalition. In the aftermath of the scandal, the Germans tried to pinpoint the source of the leak. Although there was no question that their codes had been broken, it was assumed that either the documents had been stolen or somebody had betrayed their contents.

There was one German agent operating in Argentina who defied all Hall's best efforts to track him down. Simply called Arnold, he was, according to Edward Bell, Hall's closest confidant at the US embassy in London, 'a quiet nice fellow who ... is one of the cleverest of all the German agents in the Western hemisphere'. Arnold made it his business to attack Allied–Argentine trade: he placed explosives on merchant ships, infected grain stores with fungus and contaminated livestock – horses, mules and cattle – with anthrax and glanders bacilli.

Arnold's biological weapons were sent from Spain by U-boat to America and then smuggled to Argentina. A combination of Room 40 intelligence and Hall's network on the ground in Spain had stopped several consignments already by intercepting orders placed by the German naval attaché in Madrid, Baron von Krohn. When von Krohn's involvement in this noxious trade was revealed, he was asked to leave Spain. However, he was determined to have the last laugh. He persuaded his French mistress, Martha Regnier, to sail to Buenos Aires on a Spanish liner with a supply of anthrax-laced sugar cubes hidden in her luggage. Room 40 was aware of the plot – it intercepted a total of 40 messages sent by Arnold to Berlin – and Hall ordered a British ship, HMS *Newcastle*, to intervene, giving exact instructions about the whereabouts of the anthrax. However, HMS *Newcastle* lost Martha's ship in heavy fog and she made it to Argentina.

Arnold got quickly to work. He infected 200 mules that were bound for Europe: they all died in transit. Another shipment of 5,400 mules heading for Mesopotamia in early 1918 was similarly affected, while a cargo of horses intended for France and Italy had to be left behind. The British minister in Buenos Aires tried his best to pressure the Argentine president into giving Arnold up, but he was reluctant to act.

Finally Arnold was given his marching orders. He quietly disappeared, only to show up in Cuba with plans to ruin the sugar crop. Luckily the war ended before he could set them in motion: in November 1918, the UK had only three weeks' supply of sugar left.

The cooperation between Hall and Washington over the use of Room 40 material ground to a halt with the Irish Question. Using the confession of Joseph Dowling as an excuse – Dowling was a member of Roger Casement's Irish Brigade who'd been caught after landing on the Irish coast with orders to foment insurrection – the British arrested 500 prominent members of Sinn Fein on the night of 16–17 May 1918. At the end of the month, 69 of them were transported to the mainland and thrown in prison.

Seeking to bolster the case against them, Hall approached Edward Bell: would it be possible to have some of the messages that had passed between Irish Republicans and Berlin published in America? Bell agreed to put Hall's proposal to the President. Wilson took 11 days to respond: his answer was no. He was 'not prepared to publish these documents at this time and is not willing to publicly sanction their publication'. He had been swayed by the advice of his closest adviser, Robert Lansing, who reminded Wilson that 'the Irish situation is very delicate and anything we do to aid either side in the controversy would ... involve us in all sorts of difficulties with Irish in this country'.

Hall was not ready to give up just yet. However, an attempt to have some of the messages released to the British press failed to get off the ground. Conscious of President Wilson's objections, and deeply divided over the fate of the prisoners and Irish policy in general, the government decided to do nothing. Frustrated by his colleagues' timidity, Edward Shortt, the Chief Secretary for Ireland, visited Hall at the Admiralty and asked to see all the messages relating to republican–German plots. He was impressed, and wanted the whole lot published. Hall refused: to do so would jeopardise Room 40's anonymity.

As it was, the prisoners never came to trial. After Sinn Fein did exceptionally well at the December 1918 elections, pressure grew for their release. But the government continued to prevaricate. It took the escape of two high-profile prisoners from Lincoln Jail at the end of February 1919 to force the government's hand. A rolling programme of releases began on 4 March: the prisoners were finally free to return to Ireland and civil war.

# Chapter 16

# 'SAFE FOR DEMOCRACY'
# – AMERICA'S
# INTELLIGENCE WAR

In his groundbreaking *Manual for the Solution of Military Ciphers*, published in 1916, US army Captain Parker Hitt begins by letting the student know the four essential pillars of codebreaking: 'perseverance, careful methods of analysis, intuition, luck.' Those very same principles were what it took for Ralph Van Deman, known as the 'Father of American Intelligence', to get the United States to agree to set up a military intelligence division after the USA declared war on Germany in April 1917. Now that the USA had finally entered the Great War after nearly three years of global carnage, the country whose economy had surged due to the war's industrial and financial demands had to make a greater surge in military intelligence. If it could not immediately rival the excellence of Room 40, then at least it would put the Americans in a more self-reliant intelligence position so they could do what the Allies expected them to do: bring the war to a victorious end.

The lanky, thoughtful Van Deman, whose long face reminded a colleague of a beardless Lincoln, was a graduate of Harvard and Yale, where he obtained a law degree. He was commissioned as a second lieutenant in the US army in 1891, when he was 26 years old. He enrolled in Miami (Ohio) medical school and by 1893 had earned a medical degree as well, entering the army as a surgeon. In 1895, while studying at the army's Infantry and Calvary School in Fort Leavenworth, Kansas, he met the scholarly Arthur Wagner, who convinced him of the importance of military intelligence. As a result, in 1897 Van Deman went to work for Wagner in Washington in the Military Information Division.

In 1915, Van Deman was assigned to the Army War College, which had taken command of military intelligence and then let it languish in bureaucratic dysfunction. When America declared war on Germany, Chief of Staff Major General Hugh Scott, a 64-year-old cavalry veteran of the Indian Wars with a tendency to fall asleep in cabinet meetings, revealed – at the highest level – the casual neglect the United States had shown toward controlling its own intelligence: Scott believed that if America entered the war, it would get the intelligence it needed from Britain or France, and that would be good enough.

It was not good enough, and Van Deman knew it. He was grateful for the intelligence help the British had given, and were giving, the United States, and now enlisted even more help from British intelligence to set up his own bureau, particularly Lieutenant Colonel Claude Dansey, an MI5

*Ralph Van Deman, the 'Father of American Intelligence'*

operative who had arrived in Washington in mid April 1917 on a British mission led by Foreign Secretary Arthur Balfour.

Dansey was part of Vernon Kell's MI5 crew, and before the war had worked as secretary of the Sleepy Hollow Country Club on Long Island. Despite its anodyne name, the club was a meeting place for some of the titans of American politics and business. One of them was Thomas Fortune Ryan, a tycoon with interests in tobacco, mass transit, insurance, coal mines and the Thompson sub-machine gun. Ryan, who was a player in New York's powerful political machine known as Tammany Hall, had been a delegate at the Democratic National Convention in 1912 that nominated Woodrow Wilson as the Democrats' man for the White House.

Van Deman exploited Dansey's powerful connections, as well as those of the police chief of Washington DC and a mysterious female novelist who had the ear of Newton Baker, the Secretary of War, to go over General Scott's obtuse head to get a homeland intelligence service ready for the war it had to fight.

His persistence and resourcefulness worked. By May 1917 he was a colonel in charge of the Military Intelligence Section, with a staff of two officers, two civilian clerks, and revealingly humble office space on the balcony of the War College Division's library, overlooking the stacks where previous intelligence files, such as they were, were lost in the volume of data no one had bothered to track when the War College merged files with the Military Information Division. Van Deman's location belied the crucial nature of his mission, for it was nothing less than 'the supervision and control of such system of military espionage and counter-espionage as shall be established … during the continuance of the present war'.

He got more help from the British and French, who provided him with intelligence on people of suspicion within the USA. The British, via Claude Dansey, also provided organisational support, and Van Deman arranged his new unit on the British model, dividing 'positive' intelligence, or gathering information on the enemy, from 'negative' intelligence, which stopped the enemy from doing the same to you. Counter-espionage, a term created by the French, was especially important to the negative intelligence mission, and in June 1917 Van Deman set up a War Department security force, taking his civilian investigators from Tom Tunney's NYPD Neutrality and Bomb Squad.

The highly secretive force operated from a private building in Washington DC. Its cover name was the pre-Orwellian 'Personnel Improvement Bureau' – and that was what it did, screening military personnel and government employees, as well as applicants for both sectors, for signs of subversive tendencies. In July, the Military Intelligence Section opened its first field office, in New York City, commanded by Special Deputy Commissioner of the NYPD Nicholas Biddle, a Harvard-educated blueblood descended from a Philadelphia banking family, and himself a banker who had been in charge of the Astor Trust, the largest landholding in New York City. Now Colonel Biddle was in charge of a force of intelligence cops given the military rank of sergeant and the police rank of inspector, operating out of NYPD headquarters on Center Street. Van Deman created six more field offices in other major cities and ports as he worked on building US intelligence into a war-ready network.

The most significant hiring that Van Deman still had to do was find someone who could run MI-8, his cryptological section, which he recognised as a critical component of military intelligence work, and something the US had never seriously practised, a lapse that Van Deman needed to repair with urgency. The obvious choice, Captain Parker Hitt, was needed for work in France. So too were the few other officers who knew a bit about the world of cryptology. Just as Van Deman was quickly exhausting his options, rescue came via a trolley car carrying a lowly code clerk from the State Department. But there was nothing lowly about Herbert Osborne Yardley in his own mind. He too had used all of his perseverance, careful methods of analysis, intuition and luck to get this meeting with Colonel Van Deman. And he was determined to come out of it as the man who could win the code war for the USA.

At the outset of America's war Yardley was the US version of Room 40 almost by himself, but he'd had to use all his poker player's wiles to convince the army to let him into the codebreaking game in the first place. Yardley, who had joined the State Department in December 1912 as a $900-a-month code clerk, had soon proved himself to have 'cipher brains' – a gift for solving codes. He had come to Washington from small-town Indiana, having been taught telegraphy by his father, a railroad station agent. He learned about codes and their decryption on his own, after digesting the US army's only pamphlet for the solution of military ciphers – the one by fellow Indiana code genius Parker Hitt – and through his own robust initiative.

Though small and skinny, Yardley had played quarterback on his high school football team, starred in school plays, sung baritone in a quartet, and been president of his class. Popular and gregarious, he befriended clerks in other embassies, who gave him copies of their code and cipher communications, and he worked on solving them while doing his regular State Department work. He was good at mathematics and a gifted poker player, talents that combined to help him master a variety of code and cipher strategies and increase his desire to push himself further.

*President Woodrow Wilson (right) and his confidant Colonel Edward House*

When he heard that White House operative Colonel Edward House, then working on a secret peace initiative in Germany, was sending a telegram to President Wilson, Yardley made a copy as it came over the wire in the code room. He thought that solving it would be the ultimate test, for surely the President of the United States would use the most sophisticated code in the world. To his own astonished dismay, he cracked the 500-word communiqué in less than two hours. 'This message had passed over British cables and we already knew that a copy of every cable went to the Code Bureau in the British navy. Colonel House must be the Allies' best informant!' Yardley marvelled. 'Is it possible that a man sits in the White House, dreaming, picturing himself a maker of history, an international statesman, a mediator of peace, and sends his agents out with schoolboy ciphers?'

When war broke out, Yardley paid a call on the Assistant Secretary of State William Phillips to get a letter of release from the State Department. Phillips, a tall, polished, Harvard-educated patrician descended from the family that founded the Massachusetts Bay Colony, tried to kill the five-foot-five, 127-pound balding scrapper from the Midwest with patronising kindness, offering him a cigarette and a raise, but also turned down his request, telling Yardley 'the Department must function, even if there is a war'.

Yardley refused to be deterred. As he didn't go on duty until 4 p.m., he still had time to make himself appear indispensable to the War Department. He finagled his way into the Signal Corps, where an officer directed him to seek out Ralph Van Deman, whom Yardley found working with his two assistants. 'He appeared old and terribly tired but when he turned his deep eyes to me I sensed his power.'

And after Yardley had told Van Deman about his own work in cryptography, America's father of military intelligence sensed Yardley's power to the point that he scribbled an order to get him commissioned into the army as fast as possible – and no Assistant Secretary of State would stand in the way. Herbert Yardley, the college dropout who loved history, now had the chance to make it: he was soon to become the first chief of MI-8.

*

In his comic poke at America's neutrality in 'The Military Invasion of America: A Remarkable Tale of the German-Japanese Invasion in 1916' in the July 1915 edition of *Vanity Fair* magazine, no less a farceur than P. G. Wodehouse, creator of Jeeves and Wooster, conjured up the heroic American Boy Scout Clarence Chugwater, who took it upon himself to save the country from the foreign peril.

> America's defenders at this time were practically limited to the Boy
> Scouts and to a large civilian population, prepared at any moment
> to turn out for their country's sake and wave flags. A certain section
> of these, too, could sing patriotic songs. It would have been well,
> then, had the Invaders, before making too sure that America lay
> beneath their heel, stopped to reckon with Clarence Chugwater.

There was some truth in Wodehouse's psychic mockery, for the United States had been playing a massive game of catch-up since declaring war on Germany in April 1917. By the summer of that year, Van Deman was building his Military Intelligence Division and trying to counter the kind of domestic agency in-fighting that had hampered the creation of a proper US intelligence service to begin with.

The 400 agents of the Bureau of Investigation (BI), who were tasked with gathering counter-intelligence on domestic subversives, were run by A. Bruce Bielaski, a career civil servant with a law degree. The BI was in a state of war with the Secret Service, run by William Flynn, a New York Irishman who had distinguished himself by combatting the counterfeiting and extortion of Black Hand anarchists, and the rising American Mafia. The Treasury Secretary William Gibbs McAdoo, who was responsible for the Secret Service, was not amused by the in-fighting, and appealed to his father-in-law, President Woodrow Wilson, to let him end this damaging bureaucratic rivalry by establishing a new centralised intelligence agency. Wilson said no.

Bielaski, stretched in both manpower and money, happily accepted intelligence help from the 'largest company of detectives the world ever saw' – the 250,000 volunteers making up the American Protective League (APL), founded by Chicago advertising executive Albert M. Briggs to counter the domestic perfidy wrought by the Germans. Astonishingly, these amateur detectives were each given a police shield badge, which read 'American Protective League' around the edge, and directly in the centre the words 'Secret Service'. Indeed, this group of citizen vigilantes was allocated $275,000 from President Wilson's $100 million emergency war fund, due to the zeal of Attorney General Thomas Gregory, who had been special counsel to the state of Texas. Gregory made his request for funds in a secretive memo, explaining that he'd reveal more 'in person'.

Gregory called the APL a 'powerful patriotic organisation', but in reality they used America's war to advance their own reactionary social and economic views by wrapping them in the flag. They burgled residences and offices, listened in on telephone conversations, intercepted and opened mail, and illegally arrested their fellow Americans. They also went to work chasing down spies and 'slackers' – men who were evading the draft that had come with the US declaration of war – and while they didn't catch a single German agent, they did make insufficiently patriotic German-Americans kiss the US flag, and ferreted out school teachers who dared to express objectivity about the war in their classrooms and got them fired. Suddenly, the United States had gone from an officially neutral land of liberty to one where every citizen was under suspicion by private deputies of the state.

Woodrow Wilson wrote a letter to his Attorney General expressing, rather lamely, worry about the excesses of the APL, and wondering – astonishingly – 'if there is any way in which we could stop it?' He did not pursue this idea with any vigour because of the political optics of opposing a quarter of a million 'patriots', and he had his own security machine in operation: the Committee on Public Information, designed to spread the government's official line on saving democracy via 75 million pamphlets distributed across the country, as well as vibrant poster art encouraging enlistment and war bond buying. The CPI also dispatched 75,000 'Four Minute Men', fast-talking patriots who had to use 'patent facts' and 'no hymn of hate' to convince dubious Americans of the war's logical virtue.

Foreshadowing another war nearly a century later, the CPI was sanitising things of German origin by renaming them, hence German measles became 'Liberty measles', sauerkraut became 'Liberty cabbage', and German shepherds became 'police dogs'. In Cincinnati, Ohio, pretzels were removed from saloon counters lest they infect beer-drinking patrons with German ideas of sedition. Municipal judges frequently fined people who failed to stand for the US national anthem at public events, and movie producer Robert Goldstein was sentenced to ten years in prison for portraying the British in an unflattering light in *The Spirit of '76*, his 1917 film about the Revolutionary War. Woodrow Wilson later commuted his sentence to three years.

With the 15 June passage of the Espionage Act, any kind of negative interference with the American military and support of the enemy – among other things – was punishable by 30 years in prison, or death. And sedition was much on the American mind in the summer of 1917, for finally American justice was going to be brought to bear on the enemy, in a San Francisco courtroom, thanks in great part to the efforts of British intelligence helping their American colleagues piece together a massive threat to the British Empire – one born in the USA

On 9 January 1915, Captain Hans Tauscher, the Krupp Industries man in New York, and a devoted operative of Franz von Papen, had set in motion a shipment of arms designed to do nothing less than free India from British rule – or exhaust Britain's army in trying to stop the Indian rebellion. Or both.

Tauscher shipped ten railway carloads of freight containing 8,000 rifles and 4,000,000 cartridges to San Diego shipbrokers M. Martinez and Company. At the same time, Ram Chandra, editor of the newspaper *Ghadr* – or 'revolution' – which had been founded in 1913 as the mouthpiece of the eponymous Sikh-Hindu party intending to foment a rebellion in India, was working with Franz von Bopp, Germany's consul in San Francisco, to procure a ship to sail the arms to India to kick off the revolution.

Indian nationalists were spread across the United States before the war, and British intelligence kept a watch on their movements. Robert Nathan of MI5, whose expertise lay in Indian sedition, had arrived in New York in

March 1916 to work with fellow Cambridge man William Wiseman and Norman Thwaites and the agents of British intelligence in America.

Nathan had spent the pre-war years in the Indian Civil Service, successfully fighting Bengal sedition, and gaining the reputation of a man who could break open any plot against the state. He burnished that reputation in 1915, uncovering a German plot in Switzerland to use anarchists to assassinate Allied leaders.

Nathan's mission in the USA was to stop the violent Indian nationalist movement then using the west coast of North America as its base to plot overthrow of British rule in India. Arms had recently been seized, but not yet the conspirators who hoped to use them to topple the British Raj. As with the Irish Rebellion, Britain was not only fighting a war against Germany and its allies, but also against movements within the Empire that wanted to see that empire destroyed. The United States, with its vast size and huge immigrant population, provided the perfect laboratory in which to concoct insurrection. The Indian nationalist movement was strongest among the Indian students at the University of California in Berkeley, across the bay from San Francisco. Har Dyal, a postgraduate student, had founded *Ghadr*, and when war began, the Germans took notice of him as a potentially powerful ally. The British and Americans wanted him removed from his seditious post, but before he could be deported, he fled for Berlin, where he worked with Otto Gunther von Wesendonck, the secretary in charge of the Indian Section of Germany's Foreign Office. The duo organised the Indian Independence Committee, with the German government providing ten million marks to promote Indian revolt against the British.

So, bankrolled by the German government in Berlin and with the full support of its agents in the USA, the Indian revolutionaries chartered a small ship, the *Annie Larsen*, which would sail with the arms of rebellion from San Diego to Socorro Island, just over 1,000 miles to the south-west. There it would rendezvous with the *Maverick*, a tanker ship that had been bought with German money. The plan was to hide the arms under oil in the *Maverick*'s cargo tanks in case of a sea search, and then to sail on to

Karachi, at the time part of India and a gateway to the Punjab, home to Sikh revolutionaries. The *Maverick* was to be met by friendly fishing vessels, who would take her cargo ashore, and 'if all went well, there would be a massacre of the garrison of Karachi, and hell would break loose over India'.

All did not go well, for the conspiracy had been infiltrated by British intelligence operatives – including Malcolm Reid, a 'special immigration officer' based in Vancouver, who ran informants along the Pacific Coast – and whose predecessor, William Hopkinson, had been killed by a Sikh assassin. The British also used the multiple talents of the occultist Aleister Crowley, who spoke several Indian dialects, made repeated trips to the West Coast, and who had already agitated on behalf of British intelligence as an Irish nationalist in New York since his arrival in the US in 1914.

When the *Annie Larsen* and the *Maverick* failed to rendezvous at Socorro Island, the arms-laden ship sailed north and on 1 July put in to port at Hoquiam, Washington, 200 miles south of the Canadian border. US agents, alerted by the British, seized the weapons, thereby inadvertently stopping another Indian plot arranged by Franz von Papen.

Germany's military attaché had been visiting Seattle in May, where he made contact with German agent Franz Schulenberg and learned that there was a sizeable population of restive Indians just across the Canadian border in Vancouver. He had paid Schulenberg $4,000 to buy a ton of dynamite, as well as 50 guns with Maxim silencers, to send to one 'Mr Singh' in Sumas, Washington, situated right on the Canadian border. The idea was that the Indian revolutionaries would sabotage the Canadian railway system and cripple the shipment of vital men and materiel to the war. When the Americans seized the weapons on board the *Annie Larsen*, von Papen told Schulenberg to stand down for fear of being caught and exposing the German embassy's nefarious reach.

It was Nathan who blew open the Hindu-German conspiracy, tipping off Inspector Tom Tunney of the NYPD that a person of interest was right under their noses. On 7 March 1917, Tunney's men arrested Dr Chandra Chakravarty, a Ghadr leader, at his home in Harlem. After initially pretending to be someone else, the diminutive, fiery-eyed Chakravarty proved to be a fountain of detail, providing names and dates of conspirators in the plot to overthrow the British in India.

Tunney and his detectives went through Chakravarty's papers, discovering that he had made $60,000 – or today's equivalent of $1.2 million – over two years without doing very much overt work. The 'little Hindu', as Tunney called him, said he had inherited the money from his grandfather in India, and that 'no less a personage than Rabindranath Tagore', the Indian poet, had paid him, in December 1916, $25,000 of the $45,000 due from the estate', with another $35,000 coming from a lawyer named Chatterji in March 1916.

Under further interrogation, Chakravarty revealed that the money had in fact come from Wolf von Igel, the right-hand man to Franz von Papen. 'I spoke of the poet, Tagore, because he won the Nobel prize, and I thought he would be above suspicion,' Chakravarty told Tunney, revealing that he had used the money to buy the house in Harlem, another house on 77th Street, where he planned to open a Hindu restaurant, and a farm at Hopewell Junction, 60 miles north of Harlem, as a meeting place for his fellow conspirators.

'And when he had given us valuable information, and had appeared at the trial, and had been himself convicted and had served his sentence (a short term) in jail, and the smoke had cleared away, he was the owner of three nice parcels of real estate and a comfortable income,' Tunney mused. 'Dr Chakravarty, although a failure as a Prussian agent, fared pretty well as an investor of Prussian funds.'

The Hindu-German conspiracy trial, with which Chakravarty was so helpful, began on 19 November 1917 in San Francisco. It was the largest Indian revolutionary trial ever held outside of India, and finally saw the United States taking charge of the sedition and sabotage that had been going on within its borders since the war began, though Robert Nathan's and British intelligence work and advice on legal strategy played a paramount role in its successful prosecution. Nearly 100 defendants were assembled before Judge William Van Fleet, including Franz von Bopp and his staff at the San Francisco German consulate, as well as the German consul at Honolulu, and 35 Indian students and revolutionaries, among them Chakravarty, Ram Chandra, and Ram Singh – a wealthy donor to the Ghadar party – and several members of a 'shipping group' who had been agents in the chartering and purchase of the *Annie Larsen* and the *Maverick*.

'The trial of these men was one of the most picturesque ever conducted in an American court,' remarked an observer. 'The turbaned Hindus lent an Oriental atmosphere. Among the evidence were publications in six Indian dialects, also coded messages, all of which called for constant translation by interpreters and cryptographers. Witness after witness recited his amazing story of adventure. The action shifted quickly between the three focal points, Berlin, the United States, and India, with intermediate scenes laid in Japan, China, Afghanistan, and the South Seas as witnesses laid out the vast international plot.'

One of the Americans called to testify at the trial was William Friedman, a legendary American codebreaker who would be commissioned a major in the US army in May 1918 and join the cryptanalysis department. Before he was seconded into the army, Friedman and his colleagues at wealthy eccentric George Fabyan's codebreaking compound outside Chicago had – when not trying to help Fabyan prove that Francis Bacon had written the plays of William Shakespeare – assisted Ralph Van Deman's fledgling MI-8 unit decoding intercepts given to them by the British, by using the 'frequency' method – of some words being more popular than others. Friedman had determined that the Indians were using a dictionary to encode their messages, and as he explained his methods to the San Francisco courtroom, the accused, reported an observer, 'glowered at one another. Had one of them sold out this secret?'

Tom Tunney, now a major in the army after his NYPD unit was drafted into intelligence service in December 1917, also testified as to the circumstances of Chakravarty's arrest. His policeman's eye noticed that the Indian defendants 'did not seem altogether fond of each other ... forever whispering, wagging their heads, stuffing notes down each other's necks and when the testimony of one of their number grew too truthful they squirmed and scowled. Chakravarty's life was threatened during the trial.'

But it wasn't Chakravarty's life that ended on 23 April 1918, the final day of the trial, when Judge Van Fleet adjourned to his chambers to prepare his charge to the jury. Ram Singh, seething at a perceived betrayal by Ram Chandra, pulled out an automatic pistol that he had procured during a recess and fired three bullets into Chandra, the fatal one entering his heart. United States Marshal James Holohan, sitting by the jury box and a 'man of great stature … shot once with his arm high over his head, so that the bullet should clear nearby counsel. The shot broke Ram Singh's neck.'

When order was restored, 29 of the defendants were convicted, with Franz von Bopp and his vice consul receiving the stiffest sentences, of two years each in a federal penitentiary, and fines of $10,000 (or $200,000 today). The sentences of the 13 Indian revolutionaries and students ranged from 22 months to 60 days, with Chakravarty – in thanks for his help – receiving 30 days in jail and a fine of $5,000, which he could easily pay off, as Tunney noted, with one of his German-financed New York properties.

'The punishment is wholly inadequate to the crime,' said Judge Van Fleet. 'The German defendants represent a system that the civilised world cannot tolerate.' British intelligence and the nascent American intelligence forces had triumphed at home, but the war in Europe, after another year of mass slaughter and the fall of the tsar, was at a critical point. With America mobilising its forces and beginning the process of packing them off to Europe, those in charge of that system realised that it was only a matter of time before they were overwhelmed on the Western Front. The Germans' best hope of forcing a quick end to the war lay with their U-boat fleet.

# Chapter 17

# U-BOAT MENACE

The entry of the USA into the war put even more pressure on the German U-boats to deliver a knockout blow as quickly as possible. By inflicting severe losses on Britain's Atlantic trade, the Germans hoped to render it unable to feed or equip its armies, reduce the civilian population to starvation, and force it to capitulate. With Britain gone, and Russia succumbing to revolution, France would surely seek peace, making America's declaration of war irrelevant, given it would take them until 1918 to reach Europe in any numbers.

The German naval Chief of Staff calculated that 'in five months, shipping from England will be reduced by 39 per cent', with the result that 'England will not be able to stand it'. Admiral Jellicoe, commander of the fleet during the Battle of Jutland and then in charge of the Anti-Submarine Division, shared this view, warning that the Germans 'will win the war unless we can stop these losses – and stop them quickly'.

Room 40 was in the front line as the sea war reached its critical phase. Despite more regular changes of the cipher keys and call signs used by the U-boats, the codebreakers, drawing on their accumulated experience, and helped by the fact that the submarines frequently used wireless on their hunting trips, produced a steady stream of detailed information on the location and destination of the enemy. One example, taken from hundreds of similar assessments collected over the course of the campaign, serves to illustrate their pinpoint accuracy: from messages intercepted on 4 April 1917, they gleaned that 'an enemy submarine was in 49 degrees 30 North 6 degrees 46 West apparently with partially disabled engine'.

At the same time, they were receiving increasing numbers of messages thanks to the 40 wireless stations that now spanned the UK: in the first two years of the war, Room 40 got an average of 27 U-boat intercepts a month; by the end of 1917 that number had jumped to 66, a figure that also reflected the intensification of U-boat activity.

Yet for all Room 40's output, it could do precious little to physically protect the merchant ships or actually destroy the enemy. The latest U-boats were a formidable weapon: 240 feet long, weighing 820 tons, with deck guns and six torpedo tubes carrying up to 16 torpedoes. In the first few months of 1917 they wreaked havoc: during January they sank 181 ships, in February 259, in the first two weeks of April alone, 373 were condemned to the bottom of the ocean, while during May, 287 suffered the same fate.

Something had to be done. These losses were simply unsustainable. Only one in four merchant ships making the transatlantic trip survived the journey. They were going down far quicker than Britain's capacity to replace them. The situation was so grave that the institutional myopia that had kept Room 40 isolated from the rest of Naval Intelligence was finally shaken off. For the first time, it really was all hands on deck.

The driving force behind this process of integration was Blinker Hall. In May, Room 40 established links with NID's German Section (ID14), which collated all relevant intelligence from human sources such as agents and coast-watchers and included a team dedicated to interrogating captured

*A U-boat on the prowl*

U-boat personnel. Meanwhile, cooperation was established between Room 40 and the Enemy Submarine Section (E1), where information received from British and neutral sources concerning 'ships attacked and reports of sightings and attacking submarines' was analysed. Previously, E1 and Room 40 had worked completely separately, ignorant of each other's existence: as Frank Birch remarked in a memo written after the war, this rigid system 'entailed an enormous waste of time and loss of efficiency'. E1 was run by Fleet Paymaster E. W. C. Thring, whose knowledge of U-boat behaviour was second to none. His section was officially absorbed by Room 40 in the autumn.

The final piece of the jigsaw was the Admiralty chart room, where staff hovered over huge maps, tracking the movement of U-boats and their prey. At last Room 40's full potential was being realised. By July, daily U-boat situation reports were being circulated containing a summary of all the intelligence gained from the various sections.

But the most radical change in the composition of Room 40 came with the introduction of women. The amount of paperwork the existing staff had to deal with had grown to such an extent that it threatened to overwhelm them. Administrative support was desperately needed and the practice of employing young women for clerical work was well established in Edwardian society.

What we know about them comes mostly from the recollections of William F. Clarke, who joined Room 40 in March 1915. A barrister before the war, he applied to the navy hoping to see some action, but due to poor eyesight was confined to shore before Hall came calling. Though not much use as a cryptographer, Clarke became a self-appointed authority on Room 40.

At first there were some objections to females entering this all-male sanctum, but as Clarke noted, 'once the ice had been broken the number of ladies rapidly increased'. The secretaries were managed by Lady Ebba Hambro, wife of Sir Everard Hambro, from a long-established merchant banking family that came originally from Copenhagen and opened their first branch in London in 1839. A forthright, no-nonsense character, she shocked her male colleagues by smoking a large cigar at one of their annual dinners, and wrote passable poetry, including a send-up of Room 40 called 'Confidential Waste'.

Working under her was a selection of well-bred, well-connected young ladies, such as Miss Harvey, daughter of Sir Ernest Harvey, secretary of the Bank of England. Not all the women were confined to clerical tasks; a few tried their hand at codebreaking. Miss Henderson, daughter of Admiral Henderson, and Miss Hayler, who worked on Italian non-alphabetical codes, were the best of the bunch.

With the women living in such close proximity with their male colleagues, in an environment charged with tension and excitement, romantic entanglements were inevitable. Alastair Denniston met Dorothy Mary Gilliat and they were married in 1917, while Dilly's friend Frank Birch fell for his future wife Vera Gage. More disruptive was the presence of Mrs Bayley, wife of a City doctor: according to Clarke, she 'caused slight but unimportant trouble by her very attractive appearance'. With these remarks he did his best to draw a veil over the truth. Mrs Bayley conducted affairs with both W. L. Fraser and Russell Clarke, the original radio buff.

Dilly's brother thought that Dilly, owing to his inexperience, 'would be quite powerless if he was thrown together for any length of time with a normal, pretty woman' and, sure enough, he found himself falling under the spell of one of his female assistants. Not that Dilly was consciously seeking a partner: his appearance, with his 'long thin wrists stretched out from the cuffs of a uniform that hung on him like a sack', and absent-minded habit of forever misplacing his spectacles and getting them confused with his tobacco pouch, would have been enough to deter a lot of women.

Not Olive Roddam, daughter of Lieutenant Colonel Roddam, part of an aristocratic Northumberland family descended from Anglo-Saxon nobility, who became his secretary during 1917. Olive would often arrive for work to find him still soaking in his think-tank, and it was in the intimate setting of Dilly's office that love blossomed. They were married in July 1920. Lytton Strachey, still smarting from Dilly's rejection of him at Cambridge, responded to news of their wedding with this acid comment: 'Dilly Knox … is to be married to an undeniable female – poor helpless vanished thing.'

With staff numbers now at an all-time high and Room 40's intelligence finally being processed efficiently, Hall had every right to feel confident that the U-boat threat would be successfully neutralised. However, the one measure that could transform the situation in the Atlantic was still being ignored until the prime minister, David Lloyd George, deeply frustrated by the intransigence of his naval commanders, descended on the Admiralty at the end of April and gave them a piece of his mind. Why weren't they using the convoy system? Surely having warships sailing alongside a convoy of merchant ships was the best way to prevent losses?

One of the arguments against convoys was that such a large collection of vessels would make an even bigger target for the U-boats. This proved not to be the case. Of 219 Atlantic convoys that sailed between October and December, only 39 were sighted by U-boats. As Churchill observed, 'the sea is so vast that the difference between the size of the convoy and the size of a single ship shrinks in comparison to insignificance. There was in fact a very good chance of a convoy of 40 ships in close order slipping unperceived between the patrolling U-boats as there was for a single ship.'

The U-boat captain Karl Dönitz, future head of Hitler's submarine forces, echoed Churchill's comments: 'for long periods at a time, the U-boats, operating individually, would see nothing at all: and then suddenly up would loom a large concourse of ships … surrounded by a strong escort of warships of all types … the lone U-boat might well sink one or two ships, or even several, but that was a poor percentage as a whole. The convoy would steam on.'

The Admiralty had been debating this issue for months without reaching agreement. Lloyd George forced its hand. By the end of May, convoys of between 30 and 40 merchant ships, with a screen of battleships around them, were introduced. To ensure they were given the benefit of Room 40's discoveries, Hall formed a Convoy Section at NID, where 'for the first time the latest information about enemy submarines could be placed next to the data regarding convoys. Any intelligence regarding the presence of U-boats could be quickly transmitted to the convoy's commanding ship, which was always equipped with wireless. This made it possible to immediately divert a convoy from a dangerous area.

Though the decline in losses was not huge, it was significant enough to avert disaster. More importantly, the number of U-boats being sunk increased. Between 1914 and 1916, only 51 were destroyed, but in 1917 the strike rate rose to 75, while in 1918, 102 were disabled. Each time one of them went down, Hall would celebrate with a glass of rum.

This modest ritual gives us some idea of how personally Hall took the U-boat menace. He understood what was at stake: the survival of the Allies as a fighting force. As a result, he put himself, and those around him, under intense pressure to provide what was needed to win his battle with the submarines. Nobody was spared. Even his compatriot Mansfield Cumming, head of MI6, felt the heat of Hall's displeasure during 1917. In September, Hall, who according to Cumming's diary was 'decidedly seedy and irritable', complained that Cumming's organisation was 'not good'.

Cumming was understandably worried that he might lose Hall as an ally. When Hall got his knighthood shortly afterwards – in recognition of his Zimmermann telegram triumph – Cumming used his letter of congratulations to try and smooth things over: 'I owe you far more than I could ever repay and wish I could serve you better … you have my unswerving loyalty and devotion.' Still unsure whether the damage had been repaired, he visited Hall on Christmas Day. Aside from seasonal greetings, he told Hall 'that I regretted very much that he is not satisfied with our Naval information', and assured him that he was doing his best. Determined to show he meant business, he made MI6 pull out all the stops to get Hall what he wanted. The results speak for themselves. From the start of the war to October 1917, MI6 had produced 260 naval reports; in the year that followed, they produced 8,000.

Slowly but surely, the tide began to turn. The accumulated effect of Allied counter-measures was taking its toll. The Germans, hampered by bottlenecks in the production process and the fact that many of their submarines were tied up in ports for repairs and maintenance, had fewer U-boats available for action. More serious was the loss of veteran commanders: out of 400 U-boat captains, a mere 20 of them accounted for 60 per cent of the damage inflicted. These warriors of the deep were irreplaceable. The balance of power tilted in favour of the Allies.

A varied and constantly improved arsenal was deployed against the U-boats. Depth charges, loaded with 300 pounds of TNT, proved fatal if they landed within 100 feet of a U-boat. The number used doubled in 1917, and doubled again the following year. Their effect was devastating. The American admiral William Sims saw the carnage they caused up close: 'first, the depth charges exploded, causing a mushroom of water ... Immediately afterward, a second explosion was heard; this was a horrible and muffled sound coming from the deep, more powerful and terrible ... an enormous volcano of water and all kinds of debris arose from the sea ... as soon as the water subsided, great masses of heavy black oil began rising to the surface.'

New, improved mines, based on a German design, went into mass production, while the British got better at minesweeping, with spectacular results: in 1917, 128 merchant ships were lost to mines; in 1918, only ten. Room 40 played its part. The codebreakers had cracked the codes used by German minesweepers and minelayers, enabling them to not only warn against danger but also to monitor the success of British mining operations.

The hydrophone, a listening device that could pick up the sound of underwater propellers, was developed. 'Dazzle painting', where ships were decorated in bizarre patterns, was introduced to fox the enemy: nearly 3,000 ships were camouflaged this way. A large number of merchant ships, known as DAMS, were fitted with cannon: 1,749 in 1917, up to 4,203 in 1918. Meanwhile, the Channel defences were strengthened and barriers built to block and entrap the U-boats. Air support was also provided. By the end of the war, there were 285 seaplanes and flying boats, plus 272 land-based aircraft with radios, touring the skies over the Atlantic.

In the USA, a massive construction programme was launched to replace the ships being lost. The American merchant fleet in April 1917 stood at

2.75 million tonnes; by September 1918 it totalled 9.5 million. In the UK, 35,000 skilled workers were pulled out of army service and sent to the shipyards to help with repairs and the construction of new vessels.

While the Atlantic remained Room 40's priority, U-boat attacks in the Mediterranean were becoming a real problem. In the spring, Hall toured the region, visiting Rome, Malta and Alexandria, and came to the conclusion that 'the time has come when every effort must be made to assist our joint naval operations in the Mediterranean. The sinkings there ... are a matter of grave concern.' What was needed, among other things, were wireless-interception and DF stations that could track the messages sent by the U-boat hunters.

The man chosen to locate and build these listening outposts was Eric Gill, a professor of physics and fellow of Merton College, Oxford, where he'd run an experimental electronics lab. Gill had joined the army in 1914 and immediately found himself digging narrow trenches on the Isle of Wight to guard against German invasion under the watchful eye of a general who, according to Gill's witty memoirs, arrived in a large, expensive car. If the trenches failed to halt the advance of the enemy, Gill's tiny force 'were to retire to the high ground ... and conduct guerrilla warfare'. Other duties included motorcycling round the island to enforce the nightly blackout, which alarmed local poachers and the odd 'shepherd with a lantern'.

It took the War Office several months to realise that this technical and theoretical expert might be more useful elsewhere. He was invited to apply for artillery training with the Royal Engineers at Woolwich, an experience that he found completely underwhelming. After six weeks of rudimentary instruction, he was summoned 'for an extremely hush-hush business' to be conducted at Lord's Cricket Ground, which turned out to be equally frustrating.

The training there was organised in tandem with the Marconi Company and consisted of a 'rather odd crowd' who spent most of the time learning to put up wireless masts as quickly as possible. Anything more sophisticated was out of the question, as the Admiralty had 'prohibited any transmission of signals and there were none that we were allowed to receive'. Next, Gill was selected for 'special wireless intelligence work'. This sounded promising until he realised he'd be sharing a room 'with an elderly officer who was in charge of all the drainage of the British army'. After his talents were finally recognised, Gill spent time masterminding the establishment of wireless facilities in the Middle East before being sent to the Mediterranean.

His first stop was Cyprus. The British government, represented by the Colonial Office, had taken control of the island in 1878, though it remained technically part of the Ottoman Empire. At the outbreak of war with the Turks, it was promptly annexed and placed under military law. During 1916, it was decided that it would be the perfect location for a DF wireless station. Cyprus was ideally placed not only to pick up German–Turkish messages from the Middle East but also to monitor the U-boat threat to Allied shipping.

By October 1917, there were 14 Austro-Hungarian submarines and 32 German ones patrolling the Mediterranean, targeting troop convoys and supply ships vital to the Allied war effort. In April of that year, they sank 254,911 tonnes. In November, the Royal Navy lost seven vessels. During 1918, U-boat attacks in the area accounted for between a third and a quarter of all Allied ships lost.

Locating signals sent by the submarines was no easy matter. Cleland Hoy noted that 'when they were dodging about in the Ionian and Aegean Seas, they proved slippery customers'. Room 40 might supply intelligence that a U-boat was 'for instance, off Andros, but though her wireless was working all the time neither we nor the Germans themselves could say where she might be tomorrow. The hundred and one possibilities of the Grecian Archipelago defeated us.' The hope was that once the DF station was built, it would prove easier to get an accurate fix on these predators.

Gill was delighted with the posting, and thought Cyprus 'one of the most delectable spots on earth'. However, with a mixed Greek and Turkish population, neither of whom was exactly trustworthy, and an administration overly concerned not to provoke them or upset their sensibilities, Gill concluded that the island was more of 'a comic opera than anything ever seen on stage'.

The authorities wanted Gill to construct the DF station a quarter of a mile from the sea. Once it was completed, they began to worry that it might be destroyed by submarine. Their solution was to get a gun to protect it. When the artillery unit arrived, they immediately set about moving into a spacious villa. Gill sardonically recalled how 'all their energies were devoted … to making the house thoroughly comfortable and six weeks after they landed the gun was still not ready for action'. Losing patience, he intervened on the basis that the preparation of the gun's position was 'more important than the redecoration'.

Almost as soon as the gun was ready, the officer in charge, worried that German spies might discover its location, ordered the nearby town 'plunged into darkness every night'. This preventative measure did nothing to stop local fishing boats plying their trade right under the nose of the gun. The officer complained that they might get hit if it turned its attention on enemy submarines. Gill suggested they actually fire it one night to act as a deterrent; the authorities, however, were horrified by the idea, fearing that an unannounced barrage would traumatise the villagers and waste ammunition.

Unperturbed, Gill pressed his case. Finally it was agreed that a test shoot would take place. All the inhabitants of the island were warned in advance. This only served to cause the panic they were hoping to avoid, and the population promptly fled to the hills. To cap it all, when the firing took place, the recoil buried the gun in the earth.

Gill's last mission in the area was to set up a DF station on the island of Mudros, home to a small naval base manned by Royal Engineers 'who were commonly reputed to spend all their time building fresh piers for the navy to land beer on'. The staff at the local HQ thought it should be built on a hill on the far side of the island. Gill pointed out that 'in actual fact a DF station requires a fairly level bit of ground' and found a better location near the naval camp. HQ thought this was too close for comfort. Gill persisted and got his way.

Once up and running, the station operated 24 hours a day. However, at night, the island was blacked out, which presented a problem for the DF operators: how to ensure they had enough light to do their job without standing out like a diamond in a pile of coal. Their solution was to use an electric lamp covered by a shade made out of red chocolate wrapper.

Unfortunately, the first night they used it, pandemonium broke out on the British flagship moored nearby, the sailors going into frenzied preparations for action because 'showing of a red light was the alarm that a Zeppelin was on its way and some enterprising sentry had seen our chocolate paper'.

Despite the absurdities encountered by Gill, his work paid dividends as the volume of intercepts available for analysis by Room 40 increased daily. Nevertheless, Hall felt more could be done. Up to this point, he had jealously guarded his codebreakers and their secrets from the other Allies. But the U-boat crisis demanded a more collaborative approach. The French Admiralty, who had overall responsibility for the Mediterranean, were given the privilege of seeing Room 40's intelligence summaries, carried by messenger to Paris every night.

Next Hall turned his attention to Italy. Its cryptographic operation was primitive by comparison, so he decided to send one of his star codebreakers, Nigel de Grey, to assist them. Fresh from his Zimmermann telegram triumph, de Grey managed to solve the Austrian and German naval ciphers used in the region by the summer of 1917. Armed with this knowledge, and accompanied by half a dozen codebreakers and several wireless engineers, he headed for Italy.

His first move was to have a wireless interception station erected at Otranto. Two more listening posts were established in Rome and Brindisi. A series of DF stations was added to the mix. With this infrastructure in place, de Grey ran the NID in Rome until the end of the war. Supported by nine officers, four female clerks and 15 naval personnel working closely with his extremely grateful Italian counterparts, he masterminded a dramatic improvement in their intelligence-gathering, limiting the U-boats' freedom of movement and strike capability.

The increase in U-boat casualties offered Hall the chance to ease the burden on his codebreakers. During 1917, the Germans had changed the SKM, their main naval code book. Without a physical copy of the new codes, Room 40 had to start from scratch. The job was made somewhat easier by the U-boat wireless operators. Faced with a more complex code, which was therefore trickier to transmit, they often, through laziness and force of habit, stuck with the old version. At the same time, Dilly Knox, who was single-handedly spending every waking moment trying to find a way into the new system, was making good progress. The codes were being painstakingly reconstructed from the ground up.

Even so, it would make a huge difference if Room 40 could obtain a copy of the book. An obvious place to look was in the carcasses of U-boats. Hall assembled a team of expert divers and gave them explicit instructions about where to find the sunken treasure: 'open torpedo hatch, open five water-tight doors, turn sharp to right and retrieve top drawer in which were all papers'.

E. C. Miller, Hall's most effective diver, explored the remains of 25 U-boats. It was a grisly business. Aside from the eerie atmosphere that haunted these ghost ships, Miller had nature's predators to contend with as they feasted on the crew's corpses: 'the dog fish are always about and will eat anything. In the mating season they naturally resent any intruder, and on lots of occasions when ... I offered them my boot ... they never failed to snap at it ... I found scores of conger eels, some seven feet long ... all busily feeding.'

On one mission, Miller recovered documents from a U-boat, sunk near Dover, which showed the Germans had broken the code used by British minesweepers. Rather than change the code, Hall decided to deceive the enemy. Knowing that it would be picked up by the Germans, he had a message sent declaring that a particular minefield off the Irish coast, which in reality was still active, had been swept clear. This false information was enough to lure a U-boat, UC-44, into the area, where it promptly hit a mine and sank. Miller then dived down to the wreck and found a copy of the new code book. Hall's gamble had paid off and Room 40 were quickly up to speed.

Overall, this unprecedented mobilisation of the manpower, technological know-how and weaponry available to the Allies shows how seriously they took the U-boat menace. At the heart of this effort was Room 40. William James, its naval representative, was not exaggerating when he wrote that 'without Room 40's information the defeat of the U-boat would have been more difficult and more prolonged'.

## Chapter 18

# DESERT CODES

During 1917–18, U-boat attacks in the Mediterranean took on special significance for the British army. Not only were much-needed troops and supplies – funnelled up the Suez Canal from the Indian Ocean – crossing the Mediterranean in order to reach the killing fields of the Western Front, they were also going the other way to join the forces engaged in desert warfare. The British were attempting to bring the Ottoman Empire to its knees by driving the Turks out of Palestine and completing the conquest of Mesopotamia (Iraq). Malcolm Hay's codebreakers in London, and their associates in the Middle East, provided invaluable and, at times, decisive assistance, both strategically and tactically.

One of MI1(b)'s priorities was to help reverse the disastrous failure of the British Indian army, known as D Force, in Mesopotamia. D Force was the creation of the India Office, which ran the subcontinent and its military policy. It resented any interference from London, and was engaged in a fractious rivalry with the administrations in Cairo and Khartoum. It saw the German-sponsored jihad – an attempt to rouse the Muslim masses living in the British Empire that was backed by the Turks and combined propaganda, sabotage and guerrilla warfare – as a grave threat to the Raj, and decided a show of strength was required to neuter it. Initially the goal was limited to securing the Persian Gulf, where Britain already had a coastal presence, thereby protecting the trade route to India and the oil recently discovered there.

Landing on 6 November 1914, D Force occupied Basra on the 21st. Moving slowly up the narrow, mosquito-infested Tigris River, they engaged

Turkish troops at Qurna and won easily. By the summer of 1915, when campaigning was suspended because of the heat, the army's ultimate objective was still not clear. On 23 October, after General Townshend had deployed near the city of Kut, the Cabinet in London approved an advance on Baghdad, some 100 miles away, despite the fact that D Force's supply lines were already stretched to breaking point.

The Turks were waiting upstream in fortified positions with a force twice the size of Townshend's. The two sides clashed at the Battle of Ctesiphon, 25 miles south-east of Baghdad. Though D Force edged the contest, it was badly bruised. Having sustained 10,000 casualties so far, Townshend decided to withdraw and headed for Kut, which was quickly encircled by the Turks.

The siege of Kut lasted nearly five desperate months; food and water dwindled to nothing, disease was rife. At the end of April 1916, Townshend decided to surrender. The 3,000 British and 6,000 Indian soldiers taken prisoner were marched hundreds of miles and deposited in camps where the conditions were appalling: few survived the ordeal. Townshend saw out the war in relative comfort near Constantinople.

The fall of Kut was a serious blow to British prestige. This humiliating defeat coincided with a significant shift in attitude in London; previously the British authorities had been reluctant to commit too many men or resources to the Middle East, but David Lloyd George, the leader of a new coalition government, was keen to extend British influence in the region.

Even before the war, many considered the Ottoman Empire to be on its knees: all it would take was a gentle nudge to topple it over. The strength and determination of the Turkish resistance during the tortuous, ill-fated Gallipoli expedition seemed to suggest otherwise. Yet the idea persisted that the Ottoman Empire would not survive the war.

This concentrated minds on the fate of the non-Turkish areas of that empire, and who would control them. Both British and French imperialists had ambitions in that direction. France coveted Syria and the Lebanon. Aside from Mesopotamia, Britain had its eyes on Arabia and Palestine. Their respective spheres of influence were formalised in the secret Sykes-Picot agreement which was signed in May 1916. The Egyptian Expeditionary Force (EEF), based in Cairo, was to be made ready to launch an offensive with Palestine as its primary target.

Up to then, its role had been defensive; its priority to protect the Suez canal, the main artery of Empire. Its only taste of action thus far had come in early 1915, when a modest Turkish force, having crossed the Sinai Desert unscathed, attempted to seize the Suez Canal. However, they were no match for the entrenched guns of the EEF. Though Turkish commanders still nursed a desire to try again, this did not translate into action. Now that a major campaign was being planned, good wireless communciations were required and Eric Gill, the wireless wizard, was the right man for the job.

Gill arrived in Cairo and after spending several fruitless days scouting for a location to erect a wireless mast, he ended up taking a trip to see the Pyramids. Once he caught sight of these ancient monuments to the glory of the Pharaohs, he realised he'd found what he was looking for. The peak of the Great Pyramid was the perfect spot for a wireless receiver/transmitter. Once attached, it had an impressive range; as Gill noted, 'the aerial turned out to be extremely efficient. With a single valve receiver a Zeppelin over England was heard by it.'

Duties accomplished, Gill remained in Cairo, where, for lack of anything better to do, he and a colleague began trying to tackle some of the messages coming in. Given that British intelligence in Egypt did not yet have a dedicated cryptographic section, all intercepts were being sent to Malcolm Hay's team at MI1(b) in London, which was making good progress with the enemy's Middle Eastern codes. Hay decided it was time to put the Egyptian operation on a proper footing. An expert was sent from London, identified by Gill in his memoirs as simply 'Mr Inductance', to train up and advise the locals. This mystery man was none other than Oliver Strachey.

When Strachey arrived in Cairo, the city had a reputation as a party town. Boozy expats and colonial staff combined their English pastimes with more exotic pursuits. At night, the streets were filled with increasing numbers of rowdy and riotously drunk soldiers, many of them Australian.

Though Strachey didn't have much in common with either of these groups he would have found ways to enjoy himself.

Strachey had no problem imparting his wisdom to Gill, but was driven to distraction by an officer who'd arrived from HQ in Salonika, which was jealous of the special attention given to Cairo. This officer knew no German at all, worked in accounts and had 'never taken any interest in ciphers, puzzles, crosswords … or anything of that nature'. He was quickly sent packing 'by the next boat'.

The knowledge Strachey brought with him had been hard won. An initial way into the German–Turkish codes was provided when a cipher used by the Turks in Yemen was found on board a captured boat. It provided useful clues about how to tackle the others. In mid 1916, MI1(b) obtained a copy of the primary Turkish code book. Lax behaviour by the German wireless chief in Constantinople provided further insights. According to Gill, after a celebration dinner to mark his return from Berlin this man sent messages of 'good cheer to all enemy stations in six different ciphers. Before this happened five of the ciphers were undecipherable by us but we knew the sixth. After … we knew the lot.' The Turks were 'even more delightful, as frequently they did not bother to encipher their messages at all', and had a 'habit of chatting to each other when business was slack'.

MI1(b) attacked the material and identified the ciphers used to foment trouble in Persia, Afghanistan and Arabia. Other ciphers dealt with tactics, the flying corps, administration and supplies. The only one that eluded them governed messages to and from the German High Command in the region: known as the Yilderim cipher, it took nearly a year to solve.

In the spring of 1917, Strachey set off for home. Near Cyprus, his ship was spotted by a U-boat and sunk by its torpedoes. Strachey, and the others who managed to get off the boat before it hit the bottom of the sea, was eventually rescued by the navy. He treated the whole incident as nothing more than a minor inconvenience.

While the EEF prepared to go into battle, steps were being taken to salvage the honour of the British army in Mesopotamia. A much larger and better

*Indian sappers lay telephone cable during the Mesopotamian campaign of 1917*

equipped force was assembled by the War Office, and a new commander, General Maude, was furnished with 150,000 men, substantially more aircraft and, for his advance up the Tigris and Euphrates rivers, 446 tugs, compared to the mere 14 available to Townshend's doomed mission.

General Maude would also benefit from more accurate intelligence about his enemy. Inadequate intelligence had dogged Townshend from the start. Although the India Office was the first government department to set up a dedicated cryptographic bureau in 1906, it concentrated on Russian codes – the tsar's empire was considered to be the most serious threat to the Raj – and did not feel the need to examine German or Turkish ones. Even if it had, any advances made would have been undermined by the lack of material to work with: in the early stages of the war, the Turks rarely used wireless, instead relying on telephone. During 1915, they laid 2,000 km of landlines while possessing only three fixed wireless stations, none of which was in Mesopotamia. This left D Force at the mercy of local agents who were often unreliable and self-serving, and had a tendency to tell the British what they wanted to hear.

Consistently lacking was accurate information on the size and composition of enemy forces. D Force overestimated the number of soldiers it
encountered in the initial actions, thus leading to overconfidence, and
underestimated them when it came to the critical showdown at Ctesiphon.
This dire situation would be rectified thanks to intelligence officers with
the necessary talents and intellectual training to successfully penetrate the
enemies' codes, aided by a powerful wireless transmitter/receiver that had
been erected in Basra, and by the fact that the Turks had by then established
half a dozen wireless stations in Mesopotamia, thereby providing the
codebreakers with enough material to work on.

The two codebreakers who helped guide Maude to victory were Gerard
Clauson and Reginald Campbell Thompson. Clauson was born in 1891
and educated at Eton, where he displayed a talent for languages. Aged 15,
he taught himself Turkish. At Oxford he studied classics and then began
an academic career, winning prizes for translations from Sanskrit while
conducting research into the Turkic family of languages, picking up
Mongolian and Tibetan along the way. During 1915, after a spell on the
Gallipoli front, he joined D Force.

Reginald Campbell Thompson was 15 years Clauson's senior and
represented a select breed of intelligence officer: the archaeologist. As a
child he collected flints and bits of Persian pottery, and by his teens was
translating ancient Assyrian texts from their original cuneiform script.
The earliest known system of writing, cuneiform was made up of symbols
carved into clay tablets. Developed by the Sumerians, it remained in use
for thousands of years, growing in sophistication as it passed through the
Babylonian, Assyrian and Hittite civilisations. The art of reconstructing
and deciphering cuneiform from archaeological fragments was the perfect
training for work as a codebreaker.

After studying Oriental languages at Cambridge, Thompson embarked
on nearly 20 years of wandering the Middle East, moving from one
excavation project to another, collecting and then recording his finds in a
series of books that included trailblazing studies of the magic, demonology

and astrology of the Babylonians. Clauson affectionately described him as a 'curious old bird with a most amazing inverted brain'. Clearly Thompson had an unusual mind. He liked to visit the cinema because 'he solved many of his hardest problems as he watched pictures floating across the screen'. But he was also a man of action; more Indiana Jones than Nutty Professor. He was a crack rifle shot and an accomplished sailor. He regarded exercise as a 'moral obligation'. He had little patience with those lacking the physical toughness to endure the rigours of field work in harsh and hostile landscapes, believing a man should be able to withstand 'heat and cold, hunger and thirst'.

In 1904, on his way to a site in Mesopotamia, he was delayed by 'a local plague' and 'some slight inconvenience caused by a rumour ... that an English doctor had been poisoning his patients'. On arrival, he discovered that 'the texts they had come to recopy were carved on the sheer face of a rock overhanging a precipice'. Thompson, suspended above the ground in a cradle made from packing cases, manoeuvred into place by ropes and pulleys, spent 16 days transcribing and photographing the inscriptions, drawing them with such a steady hand that 'hardly a tremor can be detected in any stroke or sign'.

In 1914, he was attached to the general staff of D Force, accompanying Townshend on his doomed mission. Thanks to his local knowledge and well-honed survival skills, he was able to escape the encirclement of Kut and join the relief force battling to get to the city.

Once he and Clauson had joined Maude's expedition, they set to work on the enemy ciphers. Intelligence supplied by them, and by Cairo and MI1(b), detailing the strength of Turkish forces, their position and intentions, meant Maude could organise his push on Baghdad with a clear idea of what lay ahead of him. The difference in outcome could not have been more marked. The advance began in December 1916. Kut was retaken in February 1917. Baghdad was occupied a few weeks later.

While British forces continued to push north, hoping to secure as much of this oil-rich country as possible, Clauson and Thompson settled into their 'own little room' in Baghdad. To keep fit, Thompson practised 'cut and thrust with a sabre on the roof of the mess'. Over a matter of months they solved all the enemy ciphers being used in the Mesopotamian theatre. Their efforts resulted in a steady stream of strategic and tactical information, including Turkish plans for an offensive against Baghdad, allowing Maude to take the necessary precautionary measures.

A letter Clauson wrote to MI1(b) in London provides a sense of their day-to-day existence. It starts with Clauson laid up in bed suffering from a 'sore belly' after giving 'a lifelike imitation of a village pump for three days'. Recovering, he managed to crack 'another of our local brand of cipher, the 9th of its kind', though he was concerned about 'how the Devil I can send you my notes on solving the ruddy thing' – in the end they went by submarine. He goes on to add that 'Turkish messages in new keys ... rarely present any difficulty' and that Thompson held 'the record having downed one with only 113 (letter) groups to work on'.

In November 1917, Clauson was sent to Egypt to act as chief coordinator of codebreaking in the Middle East. While there, he spent time improving and regulating the security of the Yinterim code – based on letter groups – which was used to communicate with his colleagues in MI1(b), Baghdad and Salonika.

Thompson was released from his codebreaking duties in March 1918 and proceeded to take charge of the general supervision of antiquities in Mesopotamia, penning a short history of the country between 4000 BC and 323 BC for the benefit of the British soldiers stationed there. At Ur, he dug up the remains of one of the oldest Sumerian cities with the help of Turkish prisoners, shrugging off the hostile intentions of local nomads.

Before the war, Thompson, the archaeologist turned codebreaker, had joined a major excavation of the Hittite royal palace at Carchemish in Syria under the supervision of David Hogarth. Thompson wanted his fiancée to join him on site but Hogarth wouldn't allow it. Annoyed, Thompson quit and went home to get married. The dig continued without him.

Hogarth was another leading archaeologist who played a major role in the Middle East. Like Thompson, he was an adventurer at heart, and had explored the ancient sites of the Greek islands and Asia Minor, becoming an expert on the connections that linked the civilisations. In 1909 he became keeper of the Ashmolean Museum in Oxford, overseeing its collection of artefacts. However, he found life as an administrator thoroughly boring. To escape, he organised the Carchemish expedition.

Before leaving, he offered his services to Naval Intelligence. At the time it was normal for academics and explorers dotted around the globe to carry out informal espionage duties. Hogarth was to keep watch on the German engineers engaged in constructing the Berlin–Baghdad railway. In archaeological research, as in so many other things, Germany was Britain's main rival. Its experts also doubled as spies. The Carchemish dig was visited by the archaeologist Max von Oppenheim who would go on to become the prime mover behind the German-sponsored jihad.

Though 53 at the outbreak of war, Hogarth was keen to do his bit. He joined the Royal Naval Volunteer Reserve and, after rejecting an offer to become an agent in Sofia, landed in Cairo just as a new intelligence organisation – the Arab Bureau – was being formed, to act as a 'centre to which all information on the various questions connected with the Near East will gravitate'.

Back in London, Blinker Hall was one of those lobbying for the right to oversee the Bureau's activities. He had a close relationship with Colonel Gilbert Clayton, a long-serving colonial soldier who was the director of the Arab Bureau. Unsurprisingly, Hall believed the bureau should be controlled by Naval Intelligence, though as he explained to Clayton in a letter, it would remain fairly independent: 'this seems to me the only successful plan, as you have people on the spot who are able to assess any information and its proper value'.

In the end, Hall lost out to the Foreign Office. Nevertheless, he kept in touch with Clayton, who agreed to maintain 'some unofficial channel of communication'. Hogarth, with all his years of experience and accumulated knowledge, was put in charge. As it happened, Hall knew Hogarth well, not only because he was a navy man but also because Hogarth's father was a close friend of Hall's father-in-law.

Occupying three rooms at the Savoy Hotel in Cairo, Hogarth and his staff were determined to pursue a 'forward' policy in the region, which entailed stirring up the tribes of Arabia against the Turks. The focus of this strategy was Hussein, the Sharif of Mecca, who had been made deliberately vague promises of British support if he chose to throw off the Ottoman yoke, which had tightened ever since a railroad had penetrated deep into his territory. It brought with it Turkish tax collectors, deeply resented by local tribes, and challenged the tribes' lucrative role in ensuring safe passage for the many thousands of Muslim pilgrims who made the journey to Mecca each year for the haj.

Almost immediately, the British-sponsored Arab uprising was on the brink of collapse. The British had already sunk considerable sums into Hussein's war chest and appeared to be facing a swift end to their ambitious plans to claim the desert kingdom for themselves in the name of Arab independence.

Direct military assistance was constrained by two factors: the understandable reluctance of the War Office to commit troops; and the prohibitions against non-Muslims entering the Hejaz region, and especially its most holy sites. Yet to do nothing was not an option, not least because failing to help Hussein would give encouragement to his deadly opponent, Ibn Saud, whose clan, fierce acolytes of the Wahhabi sect, a militantly fundamentalist branch of Islam, coveted his territory. What was particularly galling to the Arabists in Cairo and Khartoum was that Ibn Saud was being sponsored by their rivals for influence in the region, the India Office.

Offering a way though this tangled mess was a young archaeologist and protégé of Hogarth who had been kicking his heels as an intelligence officer in Cairo attached to the Geographical Section: T. E. Lawrence, better known as Lawrence of Arabia.

Lawrence's first great passion was for the medieval history he'd studied at Oxford. Hogarth spotted his talent and became his sponsor, securing him a four-year scholarship after graduation and employing him on the Carchemish excavation. Lawrence spent nearly three years there and found digging 'tremendous fun'. Hogarth brought him into the Arab Bureau and remained his patron throughout the war as Lawrence made his bid for glory.

Lawrence's vision offered a neat solution to the problems facing British intervention. Given that the Arab troops, such as they were, were no match for the Turkish army, he argued that they were best suited to guerrilla warfare, a style of fighting that also offered the prospect of amassing booty, a primary motivation for the Bedouins who would form the majority of the strike force Lawrence commanded.

Though Lawrence focused on the destruction and disruption of the Hejaz railway, he had an equally important task: to sever the Turkish telegraph lines between Damascus and Mecca, thereby forcing the enemy to use wireless instead. Messages sent this way could then be intercepted and decoded. Lawrence recalled that 'one of my cares was to distribute wire cutters over their rear and cut their telegraph daily'. Between March and May 1917 he mounted 15 attacks on the Turkish system; between July and October another 30, using camels to pull down the telegraph poles.

By the time the Egyptian Expeditionary Force was ready to advance into Palestine, with Lawrence's Arab irregulars acting in support, it could draw on material from more than 40 wireless and DF stations dotted across the Mediterranean. The Turks themselves were operating at least 12, greatly increasing the volume of messages available for decryption.

Despite this impressive infrastructure, its potential was still not being properly used. During the First Battle of Gaza, in March 1917, during which the EEF mounted a frontal assault on well-defended positions, the wireless station established by Gill in Cyprus intercepted a message, decoded in Cairo, which showed the Turks were on the point of buckling under the weight of sustained artillery fire.

This news was transmitted to the commander in the field, General Murray, but 'he had taken steps to see that no message could reach him, stating he would have no interference from Cairo HQ during the battle'. Having already sustained heavy casualties, and ignorant of the fact that the Turks were on the verge of collapse, he called off the attack, losing the chance of a decisive victory. At the Second Battle of Gaza, in April, the same tactics were repeated, again at high cost to men and materiel, with no advantage gained. Murray was swiftly replaced by General Allenby.

Allenby realised that what was lacking was the element of surprise. If that could be restored, a breakthrough might follow. This could be achieved by deceiving the enemy as to his real intentions. Wireless messaging offered the means to do so – not the only one, but it was the integral component that made the others work.

From decodes, Allenby knew the exact location, size, composition and tactical intent of the enemy. Armed with this knowledge, he could then decide where he wanted the Turks to think he was going to attack, and where he was actually going to. He wanted to strike inland towards Beersheba, and hoped to convince the Turks that his offensive would be directed at Gaza.

Based on the assumption that the enemy was also intercepting and decoding British communications – according to the German general Limon von Sanders, 'we often deciphered their wireless messages in spite of their frequent changes of cipher key' – disinformation could be spread. False messages were broadcast on a daily basis. To make sure they could be read, the British deliberately leaked the key to solving them by announcing it *en clair*. Whether or not the Turks had fallen for these measures could be established through their wireless chatter.

The most daring, yet simple, act of deception involved a British officer riding into an area where Turkish patrols were known to loiter with intent. Having spotted him, they opened fire and he escaped, 'accidentally' leaving his knapsack behind. In it were maps and papers laced with misleading information, plus some helpful notes on the British cipher being used to transmit the fake wireless messages. The Turks were delighted with their find, as army orders revealed: 'one of our NCO patrols … came back with some very important maps and documents … no doubt as important and valuable to the enemy as they are to us'.

Allenby launched his offensive in the autumn of 1917. Expecting the attack elsewhere, the Turks at Beersheba were taken completely unawares and were quickly overrun. With air superiority bolstered by the fact that MI1(b) had cracked the wireless code used by German aircraft, the EEF advance pushed remorselessly ahead.

Having excelled as a codebreaker on the Western Front, the journalist Ferdinand Tuohy was now roaming the Middle East as an ambassador for wireless intelligence. He recalled how the order for a Turkish assault 'was sent out by cipher by wireless from Jerusalem ... we deciphered the first message ... and were able to act ... before the enemy commander knew himself what was expected of him'. Soon afterwards, Jerusalem was in British hands; the last desperate wireless message from its Turkish defenders said it all: 'the enemy is in front of us only half an hour from here ... fighting has been going on day and night ... this is our last resistance. Adieu Jerusalem.'

As Allenby pushed towards Syria, he resorted again to deception: this time he pretended that he was going to strike inland, while his main thrust was actually directed along the coast. The Battle of Megiddo effectively ended Turkish resistance. With more troops, artillery and aircraft than his predecessor, and with Lawrence's Arabs protecting his flanks, Allenby had considerable advantages. The Turkish army was weakened by disease, malnutrition, mass desertion and the loss of its best units to a futile assault on the Caucasus. Nevertheless, victory was assured by Allenby's astute handling of the material analysed by the codebreakers in London, Cairo and Baghdad, all supplied by the wireless masts built by Eric Gill.

# Chapter 19

# CRISIS

During 1917, the war of attrition reached the home fronts. All the belligerents were suffering from shortages of food and other basic commodities. Raw materials were scarce. Rationing became the order of the day. The cost of financing the conflict was ruinous for all concerned; inflation was rampant, national debts astronomical. Strike action, labour agitation and anti-war feeling were on the increase. Those in power realised that victory or defeat would depend on who could best withstand the strain. Russia was the first to crack: it was hunger that drove the citizens of Petrograd and Moscow onto the streets to topple the tsarist regime in March 1917.

For many socialists and liberals in England who had instinctively despised the autocratic system in Russia, felt distinctly uncomfortable about any form of alliance with it, and hoped that a more democratic, progressive government might take its place, the fall of the Romanov dynasty was a cause for celebration. Oliver Strachey and his friend Leonard Woolf, husband of Virginia, honoured the revolution by opening the 1917 Club in Soho as a meeting place for sympathetic intellectuals and politicos. Ramsay MacDonald, Labour Party leader and future prime minister, was a founding member.

Not everyone was so enthusiastic about the Russian Revolution. The Allied leaders were justifiably concerned that it would lead to Russia's exit from the war, which would present the Germans with the chance to switch their forces from the Eastern Front to the West, thereby giving them numerical superiority. As it was, the aptly named provisional government, dependent as it was on British and French money to stay afloat and ward off the economic crisis afflicting Russia, stayed in the war and gambled its credibility on a new offensive against the Germans in July. It began promisingly but quickly unravelled as the Russian army disintegrated from within.

By now, many of the rank-and-file, inspired by the example of their comrades in the major cities, had joined self-governing soldiers' committees (soviets), and either rejected, changed or simply ignored orders, while officers were sacked, beaten up and sometimes hanged. At the same time, many thousands of the peasant conscripts who formed the bulk of the army melted away as they downed their weapons and drifted back to their villages, eager to take part in the spontaneous confiscation of the lands and estates of the nobility. To all intents and purposes, the Russian army had ceased to exist as a fighting force.

Prior to the chaotic events of 1917, neither Room 40 nor MI1(b) had endeavoured to break Russian codes. Malcolm Hay observed that 'no attempt was made to read Russian cables before the Revolution. These cables, enormous in number, and encoded by a complicated system, would have required special staff.' Meanwhile, over at Room 40, relations with their Russian counterparts had been extremely good, partly due to the debt of gratitude Room 40 owed the Russians for their decision to hand over the German naval code books they retrieved from the sinking *Magdeburg* early in the war. Blinker Hall and his staff liaised with Russian naval staff in Moscow, helped them with the cipher keys used by the German fleet in the Baltic, and supplied information about its movements, whereabouts and intentions: 'every step in their preparations, every movement of the squadrons was known at once ... whence the information was loyally forwarded to the Russians'. Notoriously secretive and reluctant to entertain visitors, Room 40 even opened its doors to Commander Przyleneki, a representative from Russian naval intelligence.

But, over the summer of 1917, with the German army advancing ever deeper into Russia and food shortages mounting, the country became chronically unstable; the Provisional Government's hold on power, tenuous at best, simply evaporated. Into the vaccum stepped Lenin and his

Bolsheviks. Supported by the all important workers and soldiers' Soviets that were effectively running the big cities, Lenin seized power in November and immediately declared his desire to come to terms with the Germans.

Fearing for the safety of his Russian colleagues and the sensitive material he had shared with them, Hall's immediate reaction was to cable them on 4 October: 'in view of the present situation, I earnestly beg you to burn all documents and papers concerned with our mutual work. Should situation improve I can replace everything and will keep you advised.'

But the situation didn't improve. As Lenin strengthened his grip, British policy entered a black hole of confusion and uncertainty, not helped by the swift emergence of the Bolsheviks' own deadly efficient secret service, the *Cheka*. MI6 tried to get agents, including the writers Somerset Maugham and Arthur Ransome, close to the leadership, and cooked up ever more ambitious and frankly ludicrous schemes to derail the Bolshevik juggernaut, but with little effect.

The one positive development was the arrival in London of Ernst Fetterlin, Russia's leading cryptographer, during July 1918. Born in 1873 in St Petersburg, Fetterlin studied languages at university and joined the Ministry of Foreign Affairs in 1896. Over time he became the tsar's top codebreaker, penetrating German, Austrian and British codes. With a price on his head and Russia descending into a brutal, pitiless civil war, Fetterlin, accompanied by his Swedish wife, managed to evade the Cheka and escape the country on a Swedish ship.

Totally broke, with no possessions except the clothes on his back and a large ruby ring, a present from the tsar for his services that he was considering selling, Fetterlin was welcomed with open arms by Room 40. He made an instant impact, bringing his knowledge of Austrian codes, an area so far neglected by Room 40, and displaying the kind of talent that led a colleague to remark that 'on book ciphers and anything where insight was vital he was quite the best. He was a fine linguist and he would usually get an answer no matter the language.' As for the Bolsheviks, they were slow to develop their own codes: when they did, Fetterlin was ready for them.

*

The two men who would mastermind the Bolshevik takeover, Lenin and Trotsky, were both in exile when the tsar was driven from his throne, and might not even have made it to Russia had it not been for the Germans and the British. The Germans, keen to promote chaos in their enemy's back yard, furnished Lenin with a special train to make sure he got home unmolested. Trotsky took a more circuitous route. His bizarre journey was an unforeseen consequence of yet more carnage on American soil.

On Tuesday 10 April 1917, just a few days after the United States had declared war on Germany, a series of explosions rocked 'F' Building at the Eddystone Ammunition Corporation in Chester, Pennsylvania, 17 miles down the Delaware River from Philadelphia. Working in 'F' that morning were 380 people loading shrapnel into shells with a highly explosive black powder known as the 'base charge'. 'Nearly eighteen tons of black powder, ignited in some way not yet determined, set off 10,000 shrapnel shells in the loading and inspecting building,' the *New York Times* reported, 'completely demolishing that structure and causing a series of detonations that shook a half dozen boroughs within a radius of ten miles of the munition plant.'

One hundred and thirty-three people, mostly girls and women, were killed in the deadliest act of sabotage yet staged in the United States, and initially, German agents were fingered as the likely suspects. However, the shells that exploded were part of a rush shipment to Russia, and reports soon surfaced that the plant's large contingent of Russian workers had received letters telling them to stay home on that day. An account in the *Philadelphia Inquirer* quoted a Russian inspector at the plant as saying he believed the explosion was caused by sabotage. A later report by an American investigator said: 'At the time of the explosion, a great many Russians were employed in the Eddystone plant, including a commission of inspectors, but on that day, not one of these inspectors was in the loading room where the blast occurred.'

That same investigator claimed to have intercepted a telegram to a man named Meyers in New York City, just an hour after the blast, which read: 'Explosion occurred at Eddystone. Our crowd safe. Woskoff.' The investigator thought that 'Meyers' could really be Leon Trotsky, who had arrived in New York in January and had been working for *Novyi Mir*, a socialist Russian-language newspaper. It was no secret that Trotsky opposed the war and supported German socialists as part of his dream of an international revolution; he would not want any armaments to reach the provisional government in Russia because he was planning to overthrow them. But by the time of the explosion at Eddystone, Trotsky was on his way to continue the revolution in Russia, one that had toppled tsar Nicholas's government and the tsar himself, who had abdicated on 15 March.

Trotsky was detained on his way back to the revolution thanks to a warning sent to London by Blinker Hall's naval attaché in America, Australian Guy Gaunt, who had been promoted in rank to commodore. Five days later, Trotsky, his wife Natalya, his sons Sergei, 9, and Lyova, 11, along with five other revolutionaries, had sailed from New York for Europe aboard the Norwegian-American liner SS *Kristianiafjord*. On 30 March, the ship arrived in Halifax, Nova Scotia, the closest North American port to Europe, and the assembly point for ships heading via convoy into the U-boat-infested Atlantic.

The day after the *Kristianiafjord* arrived in Halifax, a message flashed from London to the British naval control officer in Halifax, Captain O. M. Makins, instructing him to remove Trotsky and his fellow revolutionaries from the ship and await further instructions. The cable said that Trotsky and his companions were 'Russian socialists leaving [the United States] for the purpose of starting revolution against present Russian Government for which Trotsky is reported to have 10,000 dollars subscribed by socialists and Germans'. The British were worried about further destabilisation of their Russian ally which was in internal revolt but still fighting on the Allied side. Trotsky's aim and that of his ilk was to make peace with Germany, allowing the Germans to devote more men and resources to the Western Front.

Trotsky's family were lodged in Halifax, and he was carted off to an internment camp at Amherst, 100 miles north-west of Halifax, commanded by Colonel Arthur Henry Morris, whom Trotsky claimed treated him worse than the tsar's secret police. The internment camp was located in the former buildings of the Canadian Car and Foundry Company, which had been confiscated from its German owner. When Trotsky arrived in April 1917, it housed 851 German prisoners of war. Of these, about 500 were captured sailors and another 200, according to Trotsky, 'workers caught by the war in Canada', while 100 or so were German officers and 'civilian prisoners of the bourgeois class'.

Before long, the multilingual Trotsky, a spellbinding orator, had turned the camp into a mini socialist state, lecturing the prisoners about the revolution in Russia and the glories of the coming new world order, one where the government of the people would end the criminal war. The men were so in awe of him that they tried to prevent him from doing camp chores, or having to queue for food, and when Colonel Morris put him in solitary confinement after the German officers expressed alarm that he was going to turn all the Germans into communists, the prisoners responded with a petition bearing 530 signatures calling for his release.

So too did the Russians call for his release, and on 3 May Trotsky was sent on his way to join the revolution. Later that year, after helping to overthrow the provisional government, and with the Bolsheviks now in power, he became People's Commissar for Foreign Affairs and began negotiating a separate peace with Germany.

As the Russian will to fight crumbled, both sides on the Western Front were pushed close to breaking point during 1917. Though the Germans stood firm against another series of titanic offensives, the morale of their battle-weary troops began to dip dangerously low. The French army was gripped by a serious mutiny after the cataclysmic failure of yet another onslaught that promised victory but delivered only massive casualties. Order was restored, but the feeling remained that its soldiers could no longer be relied upon. For the British, many months of grinding slaughter lay ahead. By the end of the year, even the Tommies were nearing their wits' end.

All this misery was accompanied by frenetic technological innovation: anything that might provide even the most marginal tactical advantage was seized upon and hurried into action. The wireless finally came into its own, with the Germans taking the lead. The relative security of the German

*Passchendaele, 1917*

telephone system could not protect it from the realities of combat: the cables and wires it relied on were extremely vulnerable to the destructive force of artillery, and were constantly severed or mangled beyond use.

As a means of battlefield communication, the telephone's limits were clear. The brave men who ran frantically from one position to another clutching messages were a poor substitute. Seeking a better method, the Germans designed a wireless that could be operated at the front. Known as the 'trench set', it was much less cumbersome than the standard wireless sets, which needed a horse and cart to transport them around. The Allies were quick to follow suit with their own version that had a range of 2,000–3,000 yards. Soon the air was buzzing with wireless signals.

Both sides were fully aware that any messages they sent could be intercepted, and set about encoding their wireless comunications. However, if wireless was to make any meaningful tactical contribution to the ebb and flow of combat, transmission and reception had to be as instantaneous as possible. Speed was essential if a message was to have any impact at all. Having to encode then decode the signals inevitably slowed things down.

In the heat of battle, and under intense pressure, it's not surprising that operators made mistakes.

Many found the process extremely frustrating. As the Royal Engineeers' account of the British Signal service put it, 'the chief obstacle to the free use of this method of signalling was the stranglehold exercised upon it by the need for ... the use of cipher'. A senior wireless officer agreed: 'ciphers have always been the bugbear of wireless. People don't like, or they have not the time, to do the enciphering.'

As a consequence, both sides introduced codes that were relatively easy to use. The British relied mostly on the Playfair Cipher. Created in 1854 by Sir Charles Wheatstone, a pioneer of the telegraph, and popularised by his friend Lord Playfair, a fellow scientist with political influence, it substituted one pair of letters for another in the text and used a keyword inserted at the beginning of the message. The German army, which had by now established a codebreaking unit, the *Abhorchdienst*, had no problems deciphering it. Though the British looked for an alternative and tried four different versions of another cipher, none of them was foolproof.

Equally, the Allies were rarely troubled for very long by the German codes. They were contained in two books: the *Befehlstafel*, designed for trench sets and made up of bigrams of common words or expressions; and the regimental-level *Satzbüch* for whole sentences, which eventually included 4,000 mixed-letter code words. To compensate for their simplicity, the German code books were changed with increasing frequency.

At first, a new code book was issued every month, then every 15 days. MI1(b) in London dealt with some of these and managed to crack more than 30 trench codes. According to Tuohy, the 'toughest code the Germans ever evolved ... puzzled our experts for precisely three days and nights back in Cork Street'.

To try and avoid, or at least delay, the inevitable penetration of their codes, both sides tried to confuse their opponents by inserting nonsense into their messages – false or misleading information – as well as dummy groups, a random assortment of material such as quotes from popular songs, poems and proverbs mixed in with irrelevant code book entries. Though these techniques were not sophisticated enough to prevent decryption, they were often too sophisticated for the hapless wireless operators who were supposed to use them. Whether through inexperience, incomprehension or plain incompetence, errors in application persistently undermined efforts to improve cipher security.

The main challenge facing the codebreakers was the sheer volume of messages that had to be processed. In France, the numbers employed by the British for deciphering duties, though small, continued to grow, effectively trebling between 1916 and 1918. The qualities needed were outlined in a memo put together by the intelligence staff. Suitable candidates should possess 'a lively intelligence ... imagination tempered by a highly developed critical faculty', combined with 'natural flair' and 'untiring patience'.

The reflections of a British officer attached to a codebreaking unit give a picture of the men chosen to work round the clock to unlock the enemy's messages. There was a schoolmaster, a stockbroker, a solicitor's clerk and a designer of ladies' hats – 'a very rum bird' – who together occupied a 'dirty little rabbit hutch' filled with pipe smoke. They hadn't a 'scrap of discipline', and neglected to wash or shave. When a new code appeared, 'they pounced upon it like vultures on their prey' and 'would wrestle with that new problem until they had made it as clear as day'. If a code was too easy, it angered them; if it was a 'real hard nut to crack', they were in seventh heaven. Overall, there was not a single code they were not able to solve within 36 hours.

Perhaps the most talented of these codebreakers was Oswald Thomas Hitchings, who ended the war in command of the Code and Cipher Solution Section at GHQ. Before volunteering in 1914, he was a teacher. A gifted organist, he taught music at two preparatory schools before becoming a modern languages master at Bridlington Grammar School: he learnt French and German via a correspondence course offered by London University. As a musician and linguist, cryptanalysis came naturally to him. However, due to his humble background, at least relative to the social elite poring over intercepts in London, he was attached to the field censor's office in France. By chance, his commander approached him one day and asked him to take a look at some German messages. He unravelled them with ease. More followed. Hitchings was put in contact with MI1(b) and a steady stream of letters flowed back and forth as he swapped ideas and notes with one of Malcolm Hay's staff, Duncan Macgregor, a professor at Balliol College, Oxford. Their exchanges bore fruit and Macgregor was sent over to France to assist him. Hay had great respect for Hitchings, and they became firm friends.

The British interception infrastructure grew rapidly during 1917. Wireless observation groups were formed. Each one had around 75 personnel overseeing six interception and two direction-finding stations, dealing with 150–200 messages a week. All the energy and manpower invested in the operation of these new technologies produced a steady stream of knowledge. Its value, however, in terms of delivering materially different results was negligible. The appalling truth about the campaigns of 1917 was that nobody seemed to have learned anything from the horrors of 1916.

During the Third Battle of Ypres, which began in early summer and ended at Passchendaele in November, the British were supposed to break out of the salient and reach the Belgian coast, cutting deep into German-held territory. Within a few weeks, the offensive was hopelessly bogged down, yet Haig persisted in his belief that victory was within his grasp: if not, then at least the Germans were being worn down by the incessant attacks. The trouble was, so were the British.

However much historians argue about the extent to which new tactical thinking was being introduced to break the stalemate, it's hard to escape the feeling that nothing had changed. The same relentless carnage. The same stubborn adherence to the belief that one more big push would do the trick. The same indifference to the appalling suffering of the troops, sacrificed on the altar of attrition.

Had Haig, at any time from mid August, bothered to actually visit the front lines – if the flimsy, porous, mud-soaked trenches inhabited by the drenched, shivering Tommies, carved into a poisonous bog laced with shell craters filled with fetid water and rotting corpses, can even be called a front line – surely he would have seen that continuing the offensive when any realistic hope of obtaining its targets was gone was not only suicidal but criminal.

The one clear-cut strategic success of 1917 was pulled off by the Germans. Without alerting the Allies, they were able to mount a major withdrawal on the Somme front to a newly prepared defensive bulwark, dubbed the Hindenburg Line by the British, some way back from where fighting had ground to a halt at the end of 1916.

This secret move was achieved through wireless and telephone silence, misdirection, and fake preparations for an offensive at Ypres. Tuohy penned a snapshot account of the moment British interceptors, listening in on the German short-wave trench set, realised they'd been fooled. At around 10 p.m., a 'bored and jaded NCO' picked up a signal that indicated that several battalions had retired. Using direction-finding, he then tried to get a better fix on the signal, which showed the German troops were 'five miles behind where they ought to have been', leading him to conclude that 'the whole bloody German army's gone back'.

What paltry gains the British made during the Somme were now worthless and the Germans were hunkered down in a far superior position. Worse was to come. A year later, they pulled off an even greater feat of deception when they concealed the timing and location of the huge spring offensive that, having taken the Allies by surprise, almost won them the war.

# Chapter 20

# OVER THERE

On 4 July 1917, the first contingent of the American army made a triumphal entry into the European war. After a perilous journey across the Atlantic that saw the US navy repel U-boat attacks without losing a single US ship or serviceman, and sinking at least one German submarine, a battalion of the First Army of the American Expeditionary Force survived a rainstorm of flowers and adulation as they marched through the star-spangled streets of Paris, with their commander General John J. Pershing at the helm, to return a martial favour.

'*Lafayette, nous sommes ici!*' proclaimed Colonel Charles Stanton, an aide to Pershing and a fluent French speaker, as the Americans paid homage at the tomb of the Marquis de Lafayette, whose French forces had sailed to America in 1781 to help the rebels win their independence from the British.

Though the Americans were welcomed in France with a tumultuous cocktail of hope and relief, their mere presence would not be enough to declare the war an Allied victory: the collapse of Russia had eventually freed up 50 German divisions of battle-hardened Eastern Front troops to head west for one big push towards victory. Despite their muscular ambition, the Americans would not win the war by themselves.

A *New York Times* reporter, one of many covering the American Expeditionary Force's (AEF) arrival in France, met up with a French drill-master, who put the task into war-weary perspective, admiring the American troops as the very finest of 'human beings and raw material'. He issued a caution, however, one born of three and a half years of seeing soldiers march to their doom. 'But they need a deal of training. The hardest thing to teach them is not to be too brave. They must first learn to hide … Methods in this war are largely those of stealth.'

One of Pershing's soldiers was largely invisible back in the United States, but he not only became the defining face of the AEF – he and his fellow Choctaw troops would contribute to US military intelligence in a way that neither General Pershing nor Private Otis W. Leader could imagine in July of 1917. Leader had been picked out of General Pershing's 4 July parade by French artist Raymond de Warreux, who had a commission from his government to paint the ideal US soldier. The artist saw in the handsome 35-year-old Leader 'a half-blood Choctaw Indian from Oklahoma, straight as an arrow and standing over six feet tall; keen, alert, yet with calmness that betokens strength and his naturally bronzed face reflecting the spirit that they took across with them, the spirit that eventually turned the tide'.

Indeed, the unique contribution of the Choctaws would help the American army begin to turn the tide of battle against the Germans, but Pershing had many other things on his mind that summer. Pershing had come to Europe as commander of the AEF bearing the weight of a great

*Choctaw soldiers*

tragedy, as well as great and unusual advancement. In 1913, he had been appointed commander of the 8th Brigade, based at the Presidio military base in San Francisco. The brigade was deployed to Fort Bliss, New Mexico, when the Mexican revolution threatened to boil over into the US. In 1915, Pershing summoned his wife and four children to join him in New Mexico, but on 27 August 1915 he received a telegram that would have destroyed most people: there had been a fire at the Presidio, and his wife and three young daughters were dead. Only his six-year-old son had survived.

True to his training and persona, Pershing, his hair gone grey and his face lined with grief, soldiered on, drawing on his vast reserve of experience and support. Indeed, President Teddy Roosevelt had thought so highly of Pershing's military skills in the Spanish–American War that he wanted him promoted from lieutenant to colonel in 1903. Since the US army only promoted by seniority, Pershing was out of luck until Roosevelt pulled rank of his own in 1905. The President couldn't promote a captain to colonel, but he could make any man a general, and so he promoted then Captain Pershing to brigadier general, leapfrogging him three ranks, and past 832 senior officers.

Pershing had learned the value of stealth and secrecy in his hunt for Pancho Villa, and while that mission had ended in failure to achieve its objective, he brought with him in the first AEF wave many of the men who had fought in that campaign, and a profound belief in the need for a vast and deep intelligence network.

Pershing knew that the American military intelligence was at best at the back of a very sophisticated class, and at worst scorned by the Allies as yet another liability brought to the war by the still undertrained, undersupplied United States army, whose commander stubbornly insisted on keeping them together as a unit rather than loaning them to the French and British as replacement troops.

The First World War proved to be a revolutionary event for American intelligence, with the exigencies of an army in the field creating the dynamic and necessary conditions for the Americans to fully commit to the world of military intelligence. The man Pershing picked to create his field intelligence programme had served as his adjutant general in the Philippine–American War, and was twice cited for gallantry in the Spanish–American War, but at heart, Dennis E. Nolan was a football coach.

Born in 1872, the eldest of six children of Irish immigrants, Nolan grew up with the classic American trifecta of hard work, faith in God, and love of country as the keys to success. He graduated from the United States Military Academy at West Point in 1896, where his talents on both offence and defence at football earned him the coveted All-American award of excellence, and those with the bat and glove won him his varsity letter playing baseball. After serving in the field with Pershing, Nolan married the niece of American Civil War Union General Ulysses S. Grant, then returned to West Point to coach the 1902 football team to a 6-1-1 record (they beat Navy but lost to Harvard) and to teach history until being called to Washington to become part of the US army's first General Staff.

*Dennis Nolan, now a Colonel, in May 1918*

By July 1917, Pershing had followed the French model and organised the AEF staff into four equal units: G1 was Personnel, G2 Intelligence, G3 Operations and G4 Supply. The intelligence service took its prefix 'G' from the British and '2' from the French intelligence Deuxième Bureau. In the summer of 1917, Major Nolan borrowed more from the British, particularly their system for tactical support. Pershing directed that intelligence units should exist at all levels of the AEF, so Nolan created an intelligence presence from the top of the army, which at its largest in August 1918 had 1.3 million troops, down to the AEF's smallest unit, the squad, which consisted of between four and ten soldiers. At field grade, the intelligence units were known as S2.

Nolan also realised that he needed another kind of intelligence, which came down to the reality of fighting a foreign war: he needed French speakers to protect his soldiers from enemy espionage and subversion. In July, 1917, he requested 'fifty secret service men, who have had training in police work [and] who speak French fluently, be enlisted as sergeants of infantry in service in intelligence work and sent to France at an early date. As these men will be in intimate relation with the French people, it is a matter of great importance that they not only speak the language but are men of high character.'

A French-speaking officer with experience in police work was dispatched to New Orleans and New York City to find 50 French speakers who were willing to do intelligence work in France. He put advertisements in local newspapers and accepted those who could pass the army's physical examination and answer a few simple questions in French.

The first wave of men aiming to be Intelligence Police, as might be expected, were a diverse lot, and not always of 'high character'. They included a French Foreign Legion deserter (and murderer), a Russian train robber, a deposed Belgian nobleman, as well as a few French army deserters. By the end of the war, the Intelligence Police were 418 strong, and one of their group, Sergeant Peter Pasqua, a Portuguese immigrant to the USA and master of languages, became the first Intelligence Police agent to win the newly created Citation for Meritorious Service after successfully infiltrating and disrupting a group of Spanish subversives working for a German agent.

At field level, intelligence-gathering involved everything from interrogating prisoners, to interpreting aerial reconnaissance, to watching the movement of enemy units on trains – in order to parse their upcoming order of battle – to espionage, to anything else that would help command to understand what the enemy was doing on the ground. Of course, there was also all the information that passed through the air: Nolan appointed Major Frank Moorman to head G2-A6, the Radio Intelligence Section, whose job was to intercept German messages and decode them, as well as to send American communiqués as securely as possible.

Moorman was a 40-year-old blunt-speaking career army man from Michigan who had risen from private to major, and had graduated from the Army Signals School in 1915. In September 1917, Pershing moved his GHQ to Chaumont, a hilltop city in the valley of the Marne, about 150 miles south-east of Paris. It was there that the AEF created its own Room 40, though theirs was called, perhaps with irony, the Glass House, and was housed in a single-storey concrete and glass shack hidden behind one of the main barracks buildings, near the sheds that contained the commissary stores.

Inside was the eclectic brainpower of the codebreakers of G2-A6, who were divided into four sections: traffic analysis, cryptanalysis, telephone intercept of enemy air artillery spotters, and the Security Service. Enlisted men were selected not because they knew codes and how to break them, but because they knew German and had potential, displayed by their application of logic and rigour to other disciplines. One officer was a lawyer who had become an expert in archaeology; another a chess master; another was an architect who had spent years learning Hebrew, Farsi and other Oriental languages.

The codebreakers were isolated from the other men by choice and, after one or two indiscretions, by necessity. Indeed, Major Moorman had established the Security Service with four stations to monitor American radio and telephone transmissions and to report any offending breaches to prevent disaster. 'There certainly never existed on the Western Front a force more negligent in the use of their own code than was the American army,' Moorman later recalled, but then, the American army codes were often the problem. It was something that Herbert Yardley, the original MI-8 codesmith, was about to find out the hard way.

When Captain Herbert Yardley, America's leading codebreaker, arrived in London in August 1918 on the orders of General Pershing, he came with the swagger of having seen the USA's cryptographic unit, MI-8, grow from a staff of himself and two civilian employees to nearly 200 men and women working to save America from her encoded enemies. He also had his own natural swagger, one born of naiveté and robust self-belief, so he was nonplussed when the British didn't immediately open their code vault and share everything with him. 'For days I made no progress,' he recalled, but 'I consumed a great deal of tea and drank quantities of whisky and soda with various officers in the War Office. They were affable enough, and invited me to their clubs. But I received no information.' Doubly frustrating was the fact that Yardley – or the US taxpayer – was often footing the bill for this British hospitality. Yardley's dinner for himself and five British officers at the Ritz Hotel cost £5 14s 6d – or about £250 in today's currency.

Pershing had ordered Yardley to London to learn as much as he could from the British codemasters, but Blinker Hall didn't like the American, pegging him as a talkative braggart, and wouldn't let him anywhere near Room 40. Part of Yardley's problem was his need to convince the Allies that America could pull her weight in the codebreaking war, and that meant doing a sales job. So he auditioned by cracking a new British cipher destined for the front lines, and won admission to the War Office's cryptanalytic bureau, where he studied their methods, describing the experience as 'finishing my education'.

The British wanted Yardley to fly to their GHQ in France to meet the musician-linguist Captain Oswald Hitchings, whose cryptanalytic skills were said by his superiors to be worth four divisions to the British army. Colonel Ralph Van Deman, head of US military intelligence, also then in London, dissuaded him, suggesting that Yardley would have more to learn from the *Bureau de Chiffre* in France. So he went to Paris.

The French Chambre Noire was just as shuttered to Yardley as was Room 40, but his letter of introduction was good enough to get him an audience with Captain Georges Painvin, 'just recovered from a long and serious illness (the fate of most cryptographers)'. The tall, dark, 32-year-old Painvin cast his black eyes on Yardley without any interest, though Yardley would eventually win him over with his code skills and the two would become lifelong friends.

Painvin, who had graduated near the top of his class from the fabled École Polytechnique, had taught paleontology at the École Nationale Superieure des Mines in Paris, and had won first prize as a cellist at the Nantes Conservatory of Music. He had come to cryptanalysis from the

*Captain Georges Painvin, the French cryptologist*
*who broke Germany's most difficult code*

trenches after the Battle of the Marne, and was doubtless certain that this visiting American had nothing to teach him. After all, he had just saved the Allies from astonishing defeat.

In March 1918, the war was nearing its fourth anniversary. A generation of young men had been slaughtered, wounded or traumatised, Russia was in revolt, Germany was starving and restive, and her unrestricted submarine warfare had solved nothing. Now that spring was near, with the Americans massing troops in France, the Germans knew that they had to do something huge to win the war before the Americans could make a difference to the Allied cause.

The Allies expected a massive German offensive, but the problem lay in determining from where. Not only did the Germans bring up artillery in stealth; they introduced a new code, one of the most difficult in the history of cryptology. Called the five-letter AFDGX cipher (the V would come later), it was created by Colonel Fritz Nebel and his team of cryptologists, who chose the letters because they sound very different from each other when tapped out in Morse code. And they were certain it was unbreakable.

Nebel's code used the Polybius square – an ancient Greek invention consisting of a 5 by 5 grid, numbered 1–5 both horizontally and vertically, with each square of the grid filled by a single letter. Nebel's ingenious twist on this standard of cryptography was to swap out the numbers and replace them with letters. Each German letter was enciphered by two AFDGX letters, so the coded message would be twice as long as the original. The Germans then separated these pairs of letters, and scrambled them according to a key, one that changed every day. The cipher was thus in three parts: substitution, division and transposition.

When the AFDGX cipher first appeared on 5 March, it did indeed seem as if the Germans had conquered the French cryptologists. Painvin worked logically through different strategies to decipher AFDGX, failing at them all while the head of the Bureau du Chiffre, Colonel François Cartier, looked over his shoulder pessimistically observing, 'This time I don't think you'll get it.'

Painvin had to get it. Direction-finding equipment showed the Allies that the messages were flowing between the Germans' top levels of command: divisions and army corps. This meant they were the overarching messages of attack. On 21 March, at 4.40 a.m., 6,600 German guns unleashed a hellish barrage against the British Fifth Army on the Somme – and the right wing of the British Third Army further north near Cambrai, a worse thunderstorm of artillery than the British had inflicted on them when the Battle of the Somme began in 1916. Five hours later, 62 German divisions – of about 15,000 men each – surged forward along the nearly 50-mile front between Arras and La Fère, and by 5 April had penetrated 38 miles into the Allied lines.

In a war whose gains and losses were measured in yards, the profound shock struck to the core of Allied intelligence. The reeling head of the *Deuxième Bureau* at France's GHQ said: 'by virtue of my job I am the best informed man in France, and at this moment, I no longer know where the Germans are. If we're captured in an hour, it wouldn't surprise me.'

The Allies retreated to Amiens, and staunched the German assault. As the Germans requested more artillery support, they generated more messages, giving Painvin more codebreaking ammunition. Painvin worked like a man staring death in the face to crack the cipher. By the beginning of April, after working 48 hours straight, he had solved the AFDGX. But the Germans kept changing it, forcing Painvin to keep pace. By working through old messages from April and May, he was able to create the German code key they had used for their March offensive. By late May, he could use this key to break the current codes.

But the situation had moved from dire to desperate. The German stormtroopers had pummelled their way another 30 miles south, and were now just 40 miles from Paris, which they were shelling with a terrifying new weapon. The 'Paris Gun', which with its range of 75 miles and its 100-foot barrel was so large it had to be mounted on railway cars, had rained shells seemingly from out of nowhere down in front of the Gare de l'Est, by the Quai de la Seine, and in the Jardin des Tuileries. On Good Friday, one of its 234-pound shells had landed on the roof of the St-Gervais-et-St-

Protais church while a Mass was in progress, killing 88 people and wounding another 68. In all, the Paris Gun killed 250 Parisians and wounded 620, its fatal shell reaching an altitude of 25 miles at the highest point in its trajectory. It was the greatest height yet attained by a weapon of war, one requiring the gunners to calculate the rotation of the earth before they fired. For the first time in history, death came down on humanity from the stratosphere.

As if that wasn't enough pressure on Georges Painvin, on 1 June the Germans added another letter to their fiendish code. The AFDGVX cipher was now in play, and yet within 24 hours Painvin had solved it. The challenge now was to intercept the message that revealed where the next attack would be launched.

At around 9 p.m. on 1 June, the French listening post at Mont-Valérien intercepted a message that according to direction-finders had been sent from German HQ to its 18 corps. The message read: 'Rush munitions Stop Even by day if not seen.' The Germans were desperate, moving arms by daylight if they could to Remaugies, and now the Allies could plan on how to counter them there. When 15 German divisions attacked on 9 June, they only advanced six miles before being repelled by five French divisions. The French called the deciphered message – thanks to Painvin – 'La Radiogramme de la Victoire'.

Painvin's codebreaking had not won the war. But he had stopped it from being lost. And now the AEF was reaching strength and about to enter the fight – with a secret code weapon that no amount of code-logic could teach, or break.

The US army had marched off to war with three authorised code and cipher systems, none of them inspiring confidence in the Allies. The Telegraph code was a fragile system designed for headquarters communications; the Signal Corps had created a highly flammable celluloid device called the Army Cipher Disk, but it was a simplistic tool for mono-alphabetic substitution, the kind of elementary code-making that uses a fixed alphabet where letters are substituted for other letters – for example, a = z, b=y, and so on – and which can easily be solved by analysing the frequency of letters and corresponding them with the frequency of likely letters in the original language (for example, in English, the letter 'e' would appear most frequently in whatever its substituted form).

The British taught the Americans how to use the Playfair cipher, which had been in use since the mid nineteenth century and which replaces pairs of letters in plain text with pairs of letters in cipher text. While far more complicated than mono-alphabetic substitution, it too could be solved with frequency analysis, but the point was to give the Americans a crash course in code creation. The British and French also provided the Americans with obsolete code books to show them how it was done, with the result being the first American Trench Code in 1918, containing 1,600 words and phrases designed to be used with 'super-encipherment' – enciphering a message more than once. The Trench Code was used for training only, and was never sent to troops on the front lines for fear of capture.

The US quickly needed its own set of codes that could be easily used by operators in battlefield conditions, and yet which were secure enough to confound the Germans. They devised the River codes, a series of two-part codes, separated into encoding and decoding, named after great American rivers, which were issued to front-line forces in August 1918.

Despite the American soldier's casual use of *en clair* communication when speaking to colleagues on field telephones, a common offence that drove commanders and cryptanalysts to distraction, the US soon stumbled on a code so secure that not even their best cryptanalysts could break it: the mother language of the army's aboriginal soldiers.

Solomon Bond Louis was underage when he volunteered to go to war for a country that didn't even consider him a citizen – and wouldn't until 1924. As a member of the Choctaw Nation, Louis hailed from an aboriginal group that had been moved from its ancestral homes east of the Mississippi River to the flat, hot scrub of Oklahoma by the Indian Removal Act of 1830, which opened up the lush homelands of the five 'Civilised Tribes' in Louisiana, Georgia, Florida and Alabama to European settlers.

Louis attended Armstrong Academy in Bryan County, Oklahoma. Like so many soldiers of the Great War, when he saw his older friends joining up, he wanted to go with them. So he pretended to be 18 years old, and after basic training that saw the aboriginals prepare for war with sticks simulating rifles (so too did African-American soldiers), he wound up with 17 of his fellow Choctaw soldiers in the 36th Division, going to the Western Front on 6 October 1918.

Leader and Bond and their fellow Choctaw would indeed turn the tide via a dynamic intersection of old world and new. For the AEF command to communicate with troops in the front lines, the US army overwhelmingly preferred the telephone. Switchboards were built underground at division headquarters to withstand enemy shelling, and in order to release men for the front lines, the AEF recruited 200 French-speaking female telephone operators from commercial telephone companies to serve overseas as civilian members of the Signal Corps Female Telephone Operators Unit, known to all who dealt with them in their postings from Chaumont to Paris to London as the Hello Girls.

For troops in the field, phone lines were strung on four-foot stakes or run along trench walls from the switchboards to each infantry battalion, and they also linked adjoining battalions. As Colonel A. W. Bloor, commander of the 142nd Infantry Regiment later explained in a memo to the division's commander: 'The field of rocket signals is restricted to a small number of agreed signals. The runner system is slow and hazardous. TPS [*telegraphie par sol* – driving iron poles into the ground to pick up electrical currents by means of induction] is always an uncertain quantity. It may work beautifully and again, it may be entirely worthless. The available means, therefore, for the rapid and full transmission of information are the radio, buzzer and telephone, and of these the telephone was by far the superior – provided it could be used without let or hindrance – provided straight to the point information could be given.'

The Germans – like the Allies – would tap into phone lines to listen in on conversations between the front and command, and civilian soldiers, such as those conscripted into the AEF, didn't always appreciate the need to use code when transmitting messages: it was more complicated than anything they had been used to in civilian life, it took longer, and under extreme conditions, time was always of the essence.

On 8 October 1918, the 36th Division was part of a fresh Allied attack, supported by artillery fire. As they captured German positions, the Americans noticed that the Germans had left their communication lines uncovered, as if tempting the Americans to use them. So they did so in a way that the Germans – and even the Americans – had never imagined.

The origins of how the Choctaw became the US army's code-war weapon vary from a white American officer stumbling upon them conversing in their own language to someone at HQ remembering that they had aboriginals in uniform, but far more likely is the claim from the men who were there that the Choctaw themselves suggested that they might be able to help out. Despite the team of highly-educated codebreakers who pushed their formidable intellects to the edge to decode the enemy's plans, the Choctaw brought something that no codebreaker could penetrate: their ancient language. The AEF quickly agreed, and the Choctaw were soon relaying messages over the phone in their native tongue.

As the Choctaw language had not evolved with the exigencies of mechanised warfare in mind, some improvisation was necessary. 'Big gun' was used to indicate artillery, while 'little gun shoot fast' meant machine gun. The battalion numbers were indicated by one, two and three grains of corn, and the regiment itself was referred to as 'the tribe'. The Germans were utterly confounded, having no idea what this new code was, and no way – with their Eurocentric dictionaries – to break it. When a captured German asked, 'What nationality was on the phones that night?' he was told only that it was Americans who had been on the phones.

Within 72 hours of taking code command, the Choctaw had demonstrated their value in helping the Americans succeed in that October attack. The war was far from over, but thanks to another colonised group of soldiers and their field-code ingenuity, it would be soon.

The man who would lead the push to victory for the Allies was a pear-shaped insurance salesman with a high-school diploma who had embezzled regimental funds, but Sir Arthur Currie was also a fine military mind that also knew its own limitations. Which was why he believed in intelligence.

Arthur Currie had begun his life at arms in 1897, aged 22, at the lowest rank of gunner in a militia regiment in Victoria, Canada. By the time war broke out, his skill as a marksman, his absorption of military strategy and history through books and manuals and courses, and his reputation for discipline had earned him promotion to brigadier general and command of the Canadian Expeditionary Force's 2nd Brigade.

Currie went to war burdened by a crime. He had used the profits from his successful insurance business to speculate in real estate during a land boom on Canada's Pacific coast. When the boom went bust, Currie was land rich and cash poor, and so he borrowed $10,000 of regimental funds to cover his debts, on the understanding that the regiment's honorary colonel was going to underwrite the regiment to the tune of $35,000. When he did not, Currie had become an embezzler, and it wasn't until 1917 – after Canada's prime minister Robert Borden learned that the man who would lead the Canadian army to glory was one step away from jail – that his own officers loaned him the money to resolve the theft. Indeed, such was the concern about this matter at the highest levels of Canadian government that Currie almost didn't win the command, and the final months of the war could have taken a much different turn for the Allies.

Currie, a lively wit when among his officers, appeared stiff and stern to his men, a six-foot-two, overweight taskmaster whose lack of military bearing was made even more apparent by his refusal to sport a moustache. But he cared deeply for his men, and planned every attack with their lives in mind, having learned at the Battle of Festubert in 1915 what havoc could result from poor planning and poor intelligence. Sloppy strategy thanks to commanders in the rear saw Currie's brigade lose 1,200 men in the course of just a few days.

In August 1918, Arthur Currie had both a knighthood and command of the Canadian Corps, a 100,000 strong force that, until his elevation in June 1917, had always been commanded by British officers. The massive German surge of March had been repelled, but the Canadians had been stationed in quiet sectors and had avoided most of the fighting. As a result, they were fresh and well trained in their battle plan when Currie launched the assault from 10 miles east of Amiens on 8 August 1918. It would be the beginning of a period known as 'The 100 Days' but for the young nation of Canada, they took those 100 days as their own particular triumph.

The war that had been dug into a spider's web of trenches across the muddy, cratered Western Front was, in its final phase, a largely fast-moving operation that put heavy demand on fast-moving intelligence. The Canadian Corps had learned much from their French and English colleagues, and now, by necessity, they had to improvise and innovate when it came to battlefield intelligence to keep pace with their astonishing progress.

The Battle of Amiens began, appropriately, with an intelligence feint. Though under the command of General Sir Henry Rawlinson and following the battle plan, Currie had taken charge in his own effective way, keeping the attack under wraps from even his most senior officers until 29 July, and in the days leading up to its launch, dispatching Canadian medical units to Flanders, knowing that German spies would detect them and report that the Canadians were going to attack at Ypres, 110 miles to the north. He also sent two infantry battalions into the Ypres sector, and the corps' wireless section headed north as well, transmitting messages with the full intention of having them intercepted by the Germans, further confusing them as to the location of the assault.

The Canadians had established their wireless school in June 1918, to train operators for the coming push. The wireless system was complex, requiring knowledge not only of sending and receiving messages, but also of setting up, maintaining and repairing the set in battlefield conditions. As a result, only operators who were already trained in Morse code and ciphers were sent to the school, as time was of the essence.

Even so, the Canadian Corps believed in the power of the wireless to win wars. When the Battle of Passchendaele was raging around him, Canadian 'gadget king' Brigadier General Andrew McNaughton ran a test, handing over a message to be sent simultaneously by pigeon and wireless. The wireless operator had the message off into the ether in five minutes; the bird had not yet made it into the air.

By the end of the first day of the Battle of Amiens, the Canadians had penetrated an astounding nearly ten miles into German territory, though at a significant cost: 1,036 killed and 2,803 wounded. Because the attack had moved so far so fast, it was a challenge to maintain a portable telephone system. Cable had to be laid and relaid as the Canadians surged forward, with lateral communications proving difficult due to the fact that the cables could not keep pace with the advance.

The Canadian Independent Force, a reconnaissance unit, countered this problem by sending motorcycle riders with wireless sets in advance of the attack, to report on German activity and transmit back to the counter-battery officer, who could then target Canadian artillery on the Germans as a further safeguard for the attacking Canadian troops. It was dangerous work. Private A. L. Bebeau rode his motorcycle through the enemy's lines a seemingly suicidal ten times, on each occasion drawing German fire. Bebeau transmitted the locations of the German machine-gun nests, which 'were successfully dealt with by the Armoured Cars and Machine Gun Batteries following up'.

Indeed, so important was this rapid transmission of battlefield intelligence that on 10 August, when the Canadians altered their sending wavelength to avoid interfering with French wireless operators, the Canadian Corps headquarters was besieged with phone calls from the British front asking why they had closed down. All the wireless stations within the British army had been listening in to the Canadian traffic in order to obtain intelligence, since the Canadian Independent Force had penetrated so deeply into enemy territory. During the battle, more than 120 wireless messages were transmitted, providing invaluable intelligence to everyone who was listening in – except the Germans, who were too ravaged to do anything about it.

German General Erich Ludendorff, who led Germany's war effort on the Western Front, called 8 August 'the black day of the German army' in the history of the war. Three German divisions had been pulverised, and more than 5,000 troops captured. The next three months would see the Canadians, with their allies, punch through the Hindenburg Line and chase the Germans eastward. It would come at a cost – 45,800 casualties, an eighth of the entire BEF, over the 100 days, even though the Canadians formed only about 15 per cent of the combined infantry. But the success was in part due not to intercepting information, but to rapidly gathering and transmitting it, and Arthur Currie was rightly proud of the Canadian Corps' intelligence service, writing that its 'system of collecting and co-coordinating information ... could almost be categorised as perfect'.

# Chapter 21

# **FINISHING LINE**

The first people to realise that Germany was about to collapse were the Room 40 codebreakers. Though peace negotiations had begun during September, spearheaded by President Wilson, the German military and the Kaiser still hoped to preserve some of their power intact and avoid the shame of unconditional surrender. So the killing went on.

In the Atlantic, the U-boats continued to hunt their prey. On 4 October, a passenger ship was sunk with the loss of 292 lives. Six days later, another one went down, taking 176 souls with it. Furious, Wilson demanded the Germans cease their attacks. Reluctantly they agreed.

This decision did not sit well with Admiral Scheer, the key figure in the German navy since 1916. The thought that his mighty High Seas Fleet should simply accept its fate appalled him. Better to go out with a bang than a whimper. Scheer believed that the only way to save his beloved fleet was to risk its destruction. As he put it in his account of the war at sea; 'among the naval commanders the idea still held force that the navy had to demonstrate and justify its further existence. Now this could only be done through a last decisive battle with the British.'

His plan was to concentrate his U-boats in the middle of the North Sea, protected by row upon row of freshly sown mines, with a force larger than that assembled for the Battle of Jutland waiting nearby, ready to pounce. The trap was to be set by 30 October. Room 40 got wind that something was afoot on the 22nd. Other messages deciphered over the next few days seemed to confirm that a major operation was being prepared, and Admiral

Beatty, in control of the Grand Fleet at its base in Scapa Flow, was warned to be on his guard.

The man on duty during the early hours of the 30th, when Scheer planned to be at sea, was Francis Toye, a musician and journalist who had only joined Room 40 at the beginning of the year and was completely in awe of Hall, 'the most stimulating man to work for I have ever known … when, blinking incessantly, exuding vitality and confidence, he spoke to you, you felt you would do anything, anything at all, to merit his approval'.

Toye was alone on night watch – 'my senior colleague was prostrate with influenza' – when 'various signs and portents' began to come in that suggested the High Seas Fleet was actually on the move. As a new recruit, he was understandably hesitant about sharing his suspicions with the Operations Division, but at around 2 a.m., as the evidence multiplied, he summoned up the courage to inform them. The staff there reacted with 'benevolent scepticism' and insisted they would only act if Toye was sure that the German navy was preparing a full-scale attack.

This put Toye in an unenviable position: should he give the go-ahead for the Operations Division to contact Beatty and thereby set the Grand Fleet in motion? Or err on the side of caution? As he put it, 'to have the Fleet sent out was to incur an enormous responsibility … not to have it sent out, if sent out it should be, was to incur a greater responsibility still'.

Just before 4 a.m., he bit the bullet: 'I went again to Operations to tell them it was my opinion that the German Fleet was moving.' This information was immediately cabled to Beatty, who began to mobilise his forces. All Toye could do was pray he'd made the right call, as 'in the space of an hour or two England spent some half a million pounds' and 'the DNI (Hall) and … the First Lord of the Admiralty, not to mention the sea lords and a whole bevy of admirals and captains, were roused from their beds by the insistent ringing of the telephone'.

Around 8 a.m., the tone of the messages being decoded suddenly changed. Apparently the Germans weren't going anywhere after all. That afternoon, Room 40 learnt that there would be no operations for a week. The next day it intercepted similar orders regarding routine manoeuvres.

On 1 November, it decoded messages that referred to desertions and court martials. Clearly, something strange was happening. What Scheer bitterly described as 'insubordination' had broken out amongst the rank and file. German sailors were simply not prepared to be 'uselessly sacrificed'. Crew on shore leave refused to return to their ships; others gathered on decks to sing peace songs.

More ominously, the Red Flag, the international symbol of revolt, was raised and Bolshevik slogans were chanted. Months of inactivity, poor food, squalid conditions and the rigid separation between the men and their overbearing, arrogant, patrician officers had taken their toll. Now, with the example of the Russian Revolution to spur them on, a full-blown mutiny erupted. Sailors took command of their ships, made common cause with equally disenchanted soldiers and workers, formed self-governing committees (Soviets) and took control of the ports. By 1 November, the disturbances had reached the main naval base at Kiel. Then they spread to Cuxhaven, Hamburg and Bremen. The momentum was unstoppable.

While both President Wilson and Lloyd George were still unaware of the scale of the revolt, the codebreakers knew that it spelled the end of German resistance. Room 40 intercepted requests for U-boats to fire

*Disgruntled sailors in Keil march in protest, November 1918*

on any ship flying the Red Flag, and for officers to lock away all sensitive documents, including code books and cipher keys. On 5 November, they decoded a message sent to U-139 informing it that Kiel was in a state of revolution. The writing was on the wall. Anticipating victory, Room 40 overflowed with 'excited members of staff'. Hall appeared, and 'as the signals were handed to him, he knew his work was done'.

Not that Hall could resist one final stunt, one last chance to mess with the enemy: he had photographs of British ships doctored so it appeared they were also flying the Red Flag. Hoping the fake pictures would encourage German sailors to follow the example of their British comrades, he got his agents to smuggle the images into all the major ports.

By 7 November, the gathering storm had swept inland: rebellion gripped Cologne, Hanover, Frankfurt, Dresden and Munich. Germany was staring revolution in the face. The government in Berlin, confronted by massive demonstrations by workers and communist agitators, tried to stem the tide. They demanded a ceasefire and the immediate abdication of the Kaiser. But he was reluctant to step down, believing his generals would stand by him. They didn't. On the 10th, he slipped over the border into Holland, and Germany became a republic. The following morning, at 11 a.m., the guns fell silent. The war to end all wars was finally over.

Under the terms of the armistice, the German navy had to present itself to the British at Scapa Flow, where its ships would be interned. At 11 a.m. on Thursday 21 November, Admiral Beatty signalled that 'the German flag will be hauled down at sunset ... and will not be hoisted again without permission'.

Appropriately enough, given Room 40's crucial role in countering the threat posed by the High Seas Fleet, one of its earliest recruits, Alastair Denniston, was present at the formal surrender. While Beatty felt only contempt for his beaten enemy and said that 'the whole German navy was not worth the life of a single English blue-jacket', Denniston could not help but feel sympathy for his vanquished foe: 'I confess I did feel sorry for the senior officers there. They had been efficient men, who had learnt their work, and made the German navy their career, and this was the end of it.'

The final act of the Kaiser's vainglorious attempt to build a navy to rival the British occurred seven months later, when the Germans, effectively imprisoned at Scapa Flow, scuttled the fleet in a smoothly executed night-time operation; the majority of their ships slid to the bottom of the sea.

With the German fleet out of action, the war was well and truly over for the staff of Room 40. Their farewell party was held on 11 December 1918. Suitable entertainment was provided by Frank Birch and Dilly Knox. Together they wrote a version of *Alice in Wonderland* set in Room 40. The short play was titled *Alice in ID25* (ID25 became the official designation for Room 40). Birch wrote the majority of the text, while Dilly contributed a number of poems that mimicked the style of Lewis Carroll's nonsense verse.

Carroll's surreal comic fantasy provided the perfect vehicle for parodying the world of Room 40. The novel, in which reality is disrupted and logic and reason are used as weapons to defeat logic and reason, related directly to the life of the codebreakers, sealed in their own universe with its own peculiar rules, where every day they confronted words whose meaning was concealed by a semantic fog. Birch and Dilly's approach was extremely faithful to Carroll's original, an act of homage to one of their favourite writers. Their script followed his narrative step by step, while Carroll's characters were transformed into Room 40 personnel.

The play begins with Alice walking down Whitehall, where she notices a scrap of paper with gibberish scrawled on it. As she reads it, 'a curious feeling came over her. She seemed to grow smaller and smaller and the people in the street began to fade away.' Suddenly, up pops the White Rabbit (Frank Aldcock, an eminent classicist), in a great hurry. Alice follows him and finds herself falling down the chute used for the pneumatic tubes into a room full of huge creatures, who are fast asleep. She is immediately accosted by one of them, who demands to know her time group; when no satisfactory answer is forthcoming, the creature decides she must be NSL (Not Logged or Sent) and dumps her in a big tin, where she dozes off.

On awakening, she resumes her search for the White Rabbit, encoun-tering a series of characters along the way, until pandemonium greets the arrival of a new cipher. Alice then precedes to the Directional Room, occupied by a forlorn Humpty Dumpty, 'not the creature I was – not since rationing came in', who has been conferring with Beatty and Jellicoe about how to sink the German fleet. Passing on, she arrives at a dark cul-de-sac, and a door bearing the sign 'Mixed Bathing'. Here she meets none other than 'Dilly the Dodo ... the queerest bird she had ever seen ... so long and so lean', with a face 'like a pang of hunger'.

Dilly the Dodo proceeds to recite his mantra, 'Greek or Latin, Latin or Greek', and shows her 'a sheet of very dirty paper on which a spider with inky feet appeared to have been crawling'. After getting into a dispute with his secretary about the location of his glasses and a missing ham sandwich, he leads Alice to the aeroplane experts, who are surrounded by 'an orchestra of typewriters'.

Suddenly the Mad Hatter appears, loaded down with an enormous tin of tobacco, and promptly demands tea. During the ensuing chaos, during which the Mad Hatter falls into a gigantic mug, Dilly the Dodo jumps up and declares that he 'must go to Room 40 and find fault with things'.

Alice pursues him but he slips away from her and she becomes embroiled in a confusing discussion with Little Man (Denniston), about codes, decodes, bi-grams and monograms. Before she can make any sense of it, he and his companions are snoring away.

Their slumbers are interrupted by the arrival of an imposing figure wearing a coat covered with little flags: it's Blinker Hall. The White Rabbit summons the courage to approach the DNI and offers him 'a piece of paper on which a lot of letters were written backwards and a lot of numbers upside down'.

Hall then demands to know where Captain James is: the Dormouse replies using classic Carrollian logic: 'his hours are 10 to 7 and 7 to 10'. Bemused, Hall asks for clarification. The Dormouse happily obliges: 'he's nearly always here at ten minutes to seven in the evening, and sometimes at seven minutes to ten in the morning'.

Before Hall can press him further, a dramatic announcement is made, 'It's demobilisation', sending the creatures into a demented panic that quickly turns into a singalong, a series of rousing verses in rhyming couplets that bring the show to a close.

As many of the codebreakers contemplated a return to civilian life, their chief, Blinker Hall, was keen to represent the Admiralty at the forthcoming peace conference at Versailles, eager to flex his muscles on a grand stage. However, his maverick style and willingness to bend the rules to breaking point ruled him out of contention for such a delicate mission. Instead, his continued poor health – the same persistent chest problems that had ended his career at sea – gave those who resented his success the excuse they needed to push him into retirement. In early 1919, Hall was replaced as head of naval intelligence by Captain Hugh Sinclair. The tributes that poured in from those who had worked with him did little to soften the blow.

Meanwhile, over at MI1(b), peace brought with it fresh challenges. As Malcolm Hay's team had already succeeded in breaking the diplomatic codes of so many countries, both friend and foe, it was ideally placed to eavesdrop on the secret chatter of the participants at the Versailles negotiations. To cope with the mass of material coming in, Hay's staff increased to over a hundred. Unfortunately, the effect of this unprecedented operation will never be known. No documentary evidence has survived: Britain's relationship with its Allies would have been in tatters if the truth had come out.

When Malcolm Hay took over at MI1(b), he did so only because he was guaranteed complete autonomy; the last thing he wanted was the mandarins at the War Office sticking their noses in. During the war he managed to keep outsiders at bay, but by the spring of 1919 he was fighting a losing battle. The new head of military intelligence, Sir William Thwaites, had 'no previous experience of Intelligence work and no obvious qualifications for the position'. Hay thought he was totally unsuitable for the job.

To add insult to injury, Thwaites decided that MI1(b) would no longer exist in splendid isolation: it was time the army took charge. Hay described how one afternoon, Thwaites invaded MI1(b)'s premises at Cork Street:

'he made no appointment, and we had no warning of his intended visit. I returned to my office … and found my room full of staff officers.'

Rather than quit, Hay decided to bite his lip. Aware that the creation of a new codebreaking organisation, the Government Code & Cypher School (GC&CS), was under discussion, he hung on in the hope that he might be chosen to run it. The Secret Service Committee, featuring representatives from the Admiralty, War Office, Foreign Office and the Treasury, agreed that Room 40 and MI1(b) should be amalgamated to form GC&CS but was divided over who should be in charge. The perennial rivalry between the Admiralty and the War Office once again reared its ugly head.

The argument turned on the relative achievements of the two sections. The Admiralty pointed out that the War Office only 'dealt with cables which are far more accurate than wireless'. In MI1(b)'s favour was the fact that, unlike Room 40, it had not been gifted any copies of German code books: it had reconstructed over a dozen of them without, as Hay later observed, 'outside assistance'.

To add fuel to the fire, the Admiralty insisted that it would 'only consent to pool our staff with that of the War Office on condition that Commander A. Denniston is placed in charge of the new department'. To counter this audacious move, the WO put forward Malcolm Hay as their candidate for the job. On 5 August 1919, the two men were interviewed by the committee. Hay hurt his case by refusing to work under Denniston under any circumstances, while Denniston said he was happy to serve under someone else if that was what the committee wanted.

Hay's bullish attitude cost him dear; Denniston was appointed head of GC&CS. But Hay was not alone in his poor opinion of Denniston. Even some of Denniston's Room 40 compatriots had severe doubts about his administrative skills and capacity to lead; according to William Clarke, 'many of the most capable … flatly declined to serve under Denniston', while one of them thought he was only 'fit to manage a small sweet shop in the East End'.

Nevertheless, Denniston remained in charge of GC&CS for over 20 years. Disappointed, Hay did not hang around. He retired on 21 August and was given a fond farewell and an engraved two-handled silver cup by his remaining staff. His last request was to have all the documents accumulated during his tenure destroyed. What evidence we have of MI1(b)'s activities comes from the material it shared with Room 40. The rest of it went up in smoke.

# Chapter 22

# SWANSONG

When Blinker Hall gave his farewell speech to his naval intelligence colleagues, he ended it with a warning: 'We have now to face a far, far more ruthless foe, a foe which is hydra-headed, and whose evil power will spread over the whole world, and that foe is Russia.'

Hall's fears about the spread of communism were shared not only by the rest of the intelligence community, but by most of the British establishment. Hall's response to this foe was to go into politics. For the Secret Service Committee, the question was how best to meet this threat. Should MI6 and MI5 be kept going as they were, or should a new, integrated espionage organisation take their place?

Basil Thomson, head of Special Branch and one of Hall's key allies, had his own ambitions in that direction: he wanted to control all Britain's intelligence agencies. His bid for power was fully supported by Hall; as Thomson noted in his diary (18 October 1918), 'I saw Captain Hall about secret service on a peace footing. I found him in full sympathy with my scheme for a civilian head with four departments under him, naval, military, foreign and home.'

Though his scheme failed to get off the ground, and both MI5 and MI6 survived intact – Mansfield Smith-Cumming, 'C', stayed on at MI6 until his death in June 1923 – Thomson was put in charge of a new outfit, the Directorate of Intelligence, formed in the spring of 1919. Its brief was to monitor and prevent subversive activities within the UK.

Thomson saw the directorate as a springboard: through it he hoped to achieve his original goal. However, over the years he had made some powerful enemies, who found his habit of hogging the limelight deeply distasteful. In 1921, they moved against him and presented their complaints to Lloyd George, who was still prime minister. Thomson was summarily sacked. Outraged, his loyal friend Blinker Hall, by now a Conservative MP, raised the matter in Parliament. During the course of an impassioned speech, Hall defended Thomson's reputation – 'there is no man who has been a better friend to England than Sir Basil Thomson' – and convinced 41 Tory MPs to vote against the government. It was not enough. Thomson was cast into the wilderness. Undeterred, he immediately wrote a book, *Queer People* (1922), about the foreign spies he'd dealt with during the war.

In December 1925, Thomson hit rock bottom. One evening in Hyde Park, two police constables caught him sitting next to a prostitute, Miss Thelma de Lava. According to the policemen, Thomson was in the process of 'fondling' her when they arrived. He was arrested and charged with committing an act in violation of public decency. When he reached the police station, he gave the custody officer a false name.

Desperate, he begged the police to drop the case, but to no avail. In court, he argued that he had been waiting to meet a communist party informer when Miss de Lava approached him. Not wanting to compromise his secret rendezvous, he said, he paid her a few shillings to get her to leave quietly; he denied kissing her and claimed to be adjusting his clothing when the police nabbed him. The ever-faithful Hall appeared for the defence as a character witness. The judge was not fooled: Thomson was found guilty and fined £5.

This sordid incident ended any ambitions Thomson had of being invited back into the fold. Instead, he turned his hand to crime fiction. His first effort, *Mr Pepper Investigates* (1925), featured 'the world's greatest detective, just over from America', and was followed by a series of half a dozen novels, each one a bafflingly complex murder mystery. Thomson then wrote his autobiography, *The Scene Changes* (1937). He died two years later. Naturally, Hall was there to mourn the loss of a man with whom he'd done so much to stop German agents causing havoc on both sides of the Atlantic.

*

As the dust settled on their wartime endeavours, Blinker Hall and Malcolm Hay contemplated putting their experiences down on paper. Hall went about it with his customary zeal, determined that the end product should reach as many people as possible. In 1932, on the advice of a literary agent, he hired a novelist to help co-write the book, secured a UK and US publisher, and sold serialisation rights to papers on both sides of the Atlantic.

With half a dozen chapters drafted, Hall ran into the brick wall that was the Admiralty. Having got wind of the book, it sent him a stern warning that it would 'object to any mention ... of the real names of persons and places in connection with real intelligence work and also any reference to the employment of agents in a neutral country'. On 4 August 1933, the Admiralty added intercepted messages and codebreaking to its list of no-go areas, leaving Hall with almost nothing to write about. He had no choice but to admit defeat. Thankfully, the unedited chapters were preserved and now reside with his other personal papers in the Churchill Archive, Cambridge.

Malcolm Hay did not even get this far down the road to publication. Having drafted a short introductory essay on MI1(b), he abandoned the project altogether, not wishing to violate the solemn commitment he'd made when he signed the Official Secrets Act. He even considered destroying the few notes he'd made, but luckily for posterity held on to them: they include much of what we know about the clandestine work of MI1(b).

This aborted attempt at an autobiography was not Hay's first literary effort, or his last. His first book was an account of his time as a soldier, *Wounded and Taken Prisoner*, published anonymously (1916). This was followed by a series of scholarly books on the history of Catholicism in England; one of them featured an appendix on 'Cryptology in the 17th Century', in which he cracked a previously unbroken cipher that had been used in a letter from a pro-Catholic conspirator to his masters in Rome.

During the Second World War, Hay was based at his Scottish estate and nearly became a victim of a Luftwaffe bombing raid: a stray projectile exploded so close by that a glass of water he was holding shattered in his hand and 'the windows and shutters were blown out of his room'.

Ever mindful of his own time as a POW, he got involved with the Prisoners of War Appeal, run by the Red Cross, raising funds and helping the families of bereaved servicemen. Then, as the war drew to a close, he began to consider his next book project. Profoundly shocked and horrified by the Holocaust, and furious at how such a monumental injustice could have been perpetrated so easily, he turned his ire on the Catholic Church. He felt that 'the Pope, who could have spoken as the representative of millions of Christians, was silent; silent all the time', with the result that 'Hitler took for granted that such failure to protest meant, if not approval, at least indifference'.

What began as an examination of the Vatican's passivity during the Nazi era grew into a study of its long history of anti-Semitism. The finished manuscript, *The Foot of Pride*, was so controversial that no English publisher would touch it. It did, however, win favour in America and came out there in 1950.

Its release was greeted with a flood of comment and debate, and attracted the support of prominent members of the Jewish community, including the legendary physicist Albert Einstein, himself a victim of Nazi persecution. Einstein wrote a heartfelt letter to Hay in which he praised him for tackling such a divisive topic: 'there are few historians of standing willing to work on such an unpopular subject as this. May I congratulate you and express to you my sincere gratitude for what you have done.'

During the 1950s, Hay visited Israel a number of times and produced several more books, including another investigation into the tortured relationship between Catholicism and Judaism. In September 1962 he fell seriously ill. He died on 27 December, aged 81. A private, modest man, Malcolm Hay was one of the unsung heroes of the First World War.

Blinker Hall spent the decade after leaving naval intelligence in politics. He was elected to a Liverpool constituency as Conservative MP in March 1919; his interventions in Parliament were mostly concerned with the welfare of sailors. At the next election, having taken on the thankless task of running the party organisation during the campaign, he lost his seat along with 88 other Tory MPs: Labour won and Ramsay MacDonald became prime minister.

Hall re-entered Parliament in 1925 as MP for Eastbourne but retired from politics in 1929, after painful surgery on his jaw. At the height of the Great Depression, a calamity that he feared would spark civil war, his beloved wife dropped dead of a heart attack while playing bridge. Theirs had been a long and loving relationship and Hall was devastated, relying on his Christian faith to get him through.

His own health remained poor. His doctors advised him not to spend any more winters in England, where the damp and cold ate away at his lungs. Over the next few years he visited Australia, New Zealand, California, the Caribbean and New Orleans. When not out of the country, he was a regular feature at the Garrick Club in London, and often seen in the company of his favourite field agent, the author and adventurer A. E. W. Mason.

Bizarrely enough, Hall also cultivated a friendship with Franz von Rintelen, the Dark Invader. Von Rintelen had masterminded German espionage in America until he was caught by Hall. Perhaps their shared naval background combined with mutual respect drew them together.

But Europe was facing a new breed of warrior, one that nursed fanatical hatred. Hall viewed the rise of Hitler and Nazism with alarm. As war became inevitable, he visited the new head of naval intelligence, John Godfrey, who was eager to pick his brains; according to Godfrey, Hall 'very unobtrusively offered me full access to his enormous store of knowledge, wisdom, cunning and ingenuity'.

With Britain in the front line after the fall of France in 1940, Hall had no intention of leaving for his customary winters abroad; he would not desert his country in its hour of need, even if it risked his health. He joined the Home Guard and embarked on a remarkable correspondence with Amos Peaslee, an American lawyer who had been fighting for nearly 20 years to bring the Black Tom bombers to justice and squeeze compensation out of the German government. Hall was one of Peaslee's key assets; he provided a considerable amount of Room 40 material to help bolster Peaslee's case, and they became close.

Though no longer at the heart of the action, Hall still followed every twist and turn of the war, and his letters to Peaslee became a running commentary, full of trenchant forensic analysis of events as they unfolded. All told, they amount to hundreds of pages, so compelling and insightful that Peaslee shared them with the secretaries of the US army and navy, who handed them over to President Roosevelt.

The flow of material started to dry up during the winter of 1942–3: Hall contracted a serious lung infection, from which he would never really recover. In August 1943, an operation to rectify the damage failed and he died on 22 October, aged 73.

Hall's immense contribution to the outcome of the First World War was neatly summarised in October 1919 during an awards ceremony at Cambridge University, at which he was to receive an honorary degree. The words of John Edwin Sandy, who was delivering the tribute, said it all: 'how great then are the benefits we derived from him who … found out what the Fleets of the Germans were contriving, tracking down the size and objectives of their forces, who planted his own spies throughout all the world and thwarted those of the enemy, who passed on to our commanders the intelligence in reliance of which they finally achieved victory'.

# Conclusion

# LEGACIES

The First World War is often described as marking the end of one era and the beginning of another. It might be more accurate to say that it represented the collision between two eras, as the old and the new clashed together, sometimes harmoniously, sometimes not. Military planners struggled to adapt their thinking to accommodate new technologies that were advancing more quickly than they were.

Room 40 and MI1(b) offer a striking counter-example of a successful blend of old-world brain power and new-world scientific invention. The wireless and the telegraph yielded a goldmine of information for the codebreakers, many of whom applied knowledge gained by studying ancient cultures at ancient universities. The results provided a stunningly comprehensive account of every aspect of the war. However, the ultimate test of the value of the codebreakers' intelligence was in how it was used.

Room 40's contribution to the naval war was potentially game-changing, providing as it did an unprecedented level of detail about the activities of the German fleet. Yet when it came to the crunch, like at the Battle of Jutland, the codebreakers' insights were often wasted or misinterpreted. The rigid hierarchy, inflexibility and warring egos that characterised the Admiralty dogged the development of Room 40. It was only when the U-boat threat was at its height during 1917 that it was allowed to reach its full potential.

It was the codebreakers' role in the diplomatic and espionage war that reaped the greatest rewards; whether it concerned the USA, Ireland, Spain, North Africa or South America, the results of Room 40's work constantly undermined and hampered Germany's efforts to tilt the war in its favour. Much of this success was due to the inspired leadership of Blinker Hall, who immediately recognised the importance of codebreaking and sought to maximise Room 40's impact whenever and wherever he could.

Although in a war of this magnitude there are countless moments that might be said to have changed the course of the conflict, there are perhaps only a handful of incidents that can truly be said to have fundamentally altered the outcome: the Zimmermann telegram was one of them.

MI1(b), under the astute Malcolm Hay, also achieved notable successes in the diplomatic field, though, due to Hay's insistence that the records be destroyed, the lack of evidence makes it hard to judge the real significance of this work, especially the extent to which it influenced policy. As to MI1(b)'s contribution to the overall military effort, it's a more mixed picture. The slow uptake of wireless on the Western Front, and the fact that neither side could prevent the enemy from breaking its codes, somewhat blunted the effectiveness of the intelligence supplied by MI1(b) and its staff.

Nevertheless, by 1918, the British army had finally fully committed to wireless interception and codebreaking at an operational level and was busily issuing lengthy memoranda analysing the best ways to encode and decode messages. It was in the Middle East campaigns that the efforts of MI1(b) had the greatest impact, giving the Allies a decisive edge and the means to outwit the enemy.

Ultimately, the collaboration between MI1(b) and Room 40 created a framework on which the British could build after the war, along with a pool of amazing talent that would dominate the business of intelligence-gathering from then on, leading to the celebrated achievements of Bletchley Park during the Second World War.

During the inter-war years, the head cryptographers at the Government Code & Cypher School were Dilly Knox and Oliver Strachey. The volume of messages they handled rarely fell below the levels reached during the war, thanks to a clause in the 1920 Official Secrets Act that instructed cable companies to hand over copies of all telegrams passing through the UK 'within 10 days of dispatch or receipt'.

Both were on a salary of £500 per year, not a negligible sum but far less than they were worth. Strachey could have scaled the heights of the civil service and earned considerably more; his ease with people, affability and social connections made him an ideal candidate for promotion. However, he genuinely loved his work and was content where he was. In his early years at GC&CS, he concentrated on cracking American codes. Asides from the Soviet Union, the USA was viewed as Britain's greatest rival, challenging its supremacy at sea and around the globe.

In 1925, Strachey switched his focus to Japanese codes, particularly their naval ones: Japan had built up a formidable fleet and threatened Britain's Far East Empire. By 1928, Strachey could read their operational code. During September 1934, he broke the Japanese cipher machine used by its naval attachés.

Dilly, on the other hand, given his eccentricities and his general impatience with those not on his intellectual level – he suffered all his life from being the cleverest person in the room – was not suitable for employment in any other branch of government. Not that he lacked an alternative. The door to academia was always open. His interpretation of the poems of Herodas, based on archaeological fragments that had obsessed him before joining Room 40 was finally published in 1922. Seven years later, an edition of his translation of Herodas into English was released by the Loeb Classical Library.

This renewed commitment to his former studies reflected his growing dissatisfaction with codebreaking. Some of this impatience was related to the revelations that ended diplomatic relations between Britain and the Soviet Union during 1927: the government publicly boasted about decoding intercepted messages from Moscow that revealed its subversive intent. With the cat out of the bag, the Soviets promptly adopted an almost unbreakable cipher system.

Given that Dilly had laboured for years on Soviet material, this turn of events must have been equally infuriating and disappointing. Suddenly deprived of anything to do, and looking for a task that would test his abilities, he shifted his attention to Hungarian codes, a hitherto neglected area. Dilly knew no Hungarian whatsoever and approached the whole thing as an abstract problem. Nevertheless, he prevailed; as Denniston noted, 'Hungarian was successfully broken by Knox, but it is doubtful if the results obtained at the time justified the enormous effort on his part.'

Dilly's quest to find a challenge worthy of his talents ended when the Nazis took power in Germany and he was confronted with the ultimate codebreaking conundrum: the Enigma machine. Its 676 possible settings could generate 11,000,000 different arrangements of the alphabet. Dilly devoted his life to unravelling its mysteries. His first breakthrough occurred in April 1937, when he managed to read messages sent by the Italian Enigma machine, which then allowed him to do the same for the one used by Franco in Spain. However, the German version remained unassailable.

A trip to Paris, along with Strachey and Denniston, to meet their French counterparts led to a visit to Poland in July 1939. The Poles had acquired a German machine and broken its ciphers. After a frustrating first day, during which Dilly lost his temper, he finally got to meet their key codebreaker, Rejewski, who brought him up to speed with his discoveries. Years later, Rejewski remembered how 'Knox grasped everything very quickly, almost quick as lightning'.

Armed with this knowledge, and a replica Enigma machine, Dilly moved into Elmers Cottage in the grounds of Bletchley Park. He was joined by the brilliant Alan Turing, whose decrypting machines would pave the way for the modern computer. At this point, though, Turing was a codebreaking novice, reliant on Dilly's expertise. Dilly liked Turing and appreciated his talent, but also found him 'difficult to anchor down' and struggled 'to keep him and his ideas in some sort of order': he could have been talking about his younger self.

Utilising tried and tested methods, developed during his Room 40 days, Dilly searched for any sloppy mistakes made by the Enigma operators. As a result, on 25 October 1939, he read a German Enigma message for the first time. He celebrated by writing a pastiche of the Lewis Carroll poem *Jabberwocky*.

It was a remarkable achievement, but the strain of carrying the Enigma burden for so long – years of intense mental struggle, striving against impossible odds and for incredibly high stakes – took its toll. In 1939 he was diagnosed with cancer. A preliminary operation forestalled the inevitable, but from then on he was living on borrowed time.

During the early part of 1941, Dilly and his team, which was almost entirely female, concentrated on the German naval Enigma machine and its Italian equivalent. Though the German machine proved intractable more tangible progress was made with the Italian one. Once solved, the intelligence gained allowed the British to take a significant portion of the Italian fleet by surprise at the Battle of Matapan, and sink four of its ships. True to form, Dilly marked the occasion by writing a poem.

Arguably, the most significant contribution Dilly made to the war effort was with his friend Oliver Strachey. In 1938, Strachey was put in charge of recruiting fresh blood, including the precociously gifted young mathematician Gordon Welchman, who would become a key figure at Bletchley. Welchman remembered being 'most impressed by Oliver Strachey … he seemed to be giving an overview of the whole problem of deriving intelligence from enemy communications'.

On 11 November 1939, Strachey reached the official retirement age of 65. Given the circumstances, with Hitler master of Poland, there was no chance he was going to give up. Instead, he headed for Bletchley Park. On arrival, he occupied an old school building and set up Illicit Services Oliver Strachey (ISOS), which tackled intercepted German messages sent the old-fashioned way, by hand not machine, a method still employed by the *Abwehr*, the military and diplomatic intelligence arm of Hitler's secret state.

During March 1940, Strachey started getting material from the Radio Security Service (RSS), run by none other than Eric Gill, the First World War wireless wizard. Set up in 1939, RSS intercepted covert radio communications between Nazi spies in the UK and their controllers in Hamburg. By December, Strachey had cracked the main cipher, enabling MI5 to identify and arrest all the German agents operating in Britain.

This security coup gave birth to one of the most effective and influential espionage operations ever launched. Rather than leave the captured spies to rot in jail, the newly formed XX Committee decided they'd be far more useful as double agents. Strachey's discoveries meant that the British could exploit the *Abwehr*'s codes and ciphers to send disinformation, transmitted by the double agents, directly to the Germans. It became known as the Double-Cross system.

The fake messages went to the Nazi embassies in Lisbon and Madrid, and then on to Berlin via an Enigma machine, a type used solely by the *Abwehr*, the Gestapo and the SS. This meant that the only way the British could confirm whether Berlin actually believed the information supplied by the double agents was by breaking into this version of Enigma.

Enter Dilly Knox. Illicit Services Knox (ISK) was established, and Dilly set about identifying a route into the machine's ciphers, a process he described with a typically obtuse analogy: 'If two cows are crossing the road there must be a point when there is only one and that's what we must find.'

On 8 December 1941, Dilly succeeded. From then on, the *Abwehr* Enigma was an open book. The Double-Cross system was the central plank of the Allies' deception campaigns before both the Sicily landings and, more importantly, D-Day. Strachey and Dilly's work helped convince the Nazis that the invasion would occur at Calais, not in Normandy. Had they known the truth, things could have turned out very differently.

Not long after his *Abwehr* breakthrough, Dilly's cancer returned with a vengeance. He was dying. He returned home but refused to give up work; his Bletchley colleagues were regular visitors. As the end approached, he was awarded the Companion of the Order of St Michael and St George (CMG). Too sick to go to the palace to collect it, he had the palace come to him. His son Oliver remembered how Dilly 'insisted on dressing and sat, shivering, in front of the large log fire, as he awaited the arrival of the Palace emissary. His clothes were now far too big for him, his eyes were sunk in a grey face, but he managed.' On 27 February 1943, Dilly passed away, aged 58.

A few weeks earlier, Strachey had suffered a major heart attack. While recuperating in hospital, he received the Companion of the British Empire

(CBE). Unable to return to Bletchley, he was forced to retire. His benevolent disposition was sorely tested by the passing years, as old age strengthened its grip. He was often depressed, and drank heavily. In 1958, he contracted pernicious anaemia, thanks to his consumption of four bottles of whisky a week. Enfeebled, he went into a nursing home, dying on 14 May 1960. He was 86 years old. Of all the First World War codebreakers, Strachey had served the longest.

The performance of the codebreakers in the two world wars clearly demonstrated their indispensability. Equally it showed the importance of developing technology that enabled them to reach their full potential. The accumulated impact of the codebreakers' work changed the course of both conflicts. The achievements of the rest of the intelligence community pale by comparison.

The interception/transmission and decoding of information remained crucial for maintaining parity during the Cold War as the surveillance state grew in size – the Americans created the National Security Agency (NSA), while in the UK, GC&CS became GCHQ. The techniques used by these agencies evolved ever greater complexity, sophistication and reach: electronic bugging, spy satellites and finally the computer gave them unprecedented power in their ability to spy on friend and foe alike.

However, nowadays the technologies that were previously the monopoly of the state are available to us all. Though the surveillance capabilities of governments are awesome, anybody armed with encryption and decryption skills can return the favour.

The First World War was the crucial foundation point of the surveillance society we live in today. Our current information age rests entirely on coding: from software engineers to shadowy organisations operating out of nondescript buildings in China, from the bedroom hackers to the criminal networks the basic DNA of it all consists of encrypting and decrypting codes.

The crossword puzzle fanatics, linguists, academics, radio hams and inventors who laboured to find a way to end the terrible human tragedy that was the First World War and stop Europe from drowning in its own blood could never have anticipated where their work might lead. The thought that they were the progenitors of the surveillance society, and that their heirs would either be working at the Pentagon pinpointing suspect individuals from space, or breaking into it from their booths in internet cafés, would have seemed ridiculous, like something out of an H. G. Wells novel, and would no doubt have sent shivers down their spines.

Nevertheless, their contribution deserves to be remembered and honoured, not only because they have been largely ignored or written out of history, but also because by revisiting them we can gain a better understanding of the Great War that shaped so much of the twentieth century and still casts its shadow today.

# ACKNOWLEDGEMENTS

The authors would like to thank our agent Sonia Land, at Sheil Land Associates Ltd, for her commitment and enthusiasm, and Lucy Fawcett, also at Sheil Land, for her valuable input. We extend our gratitude to Andrew Goodfellow, at Ebury Press, for his faith in the project, to Liz Marvin, our editor, for her excellent work, and to Jon Cooksey for his expert advice. We owe a considerable debt to Mark Bentley for getting the two of us in a room together for the first time. The staff at the Churchill Archives Centre were extremely helpful and accommodating. Many thanks to Timothy Stubbs for allowing us to use material from the Hall Papers.

**James Wyllie:** Without the continued support and encouragement of my family and friends this book would have been more of a struggle and less of a pleasure to write. I hope you all enjoy the results!

**Michael McKinley:** I would like to express my deepest thanks to my wife Nancy and my daughter Rose for their inspiration, guidance and patience throughout the process of writing this book. I can't imagine how I would have done it without them.

# NOTES

## ARCHIVES
CHURCHILL ARCHIVE, CAMBRIDGE (CA)
Clarke Papers
Denniston Papers
Hall Papers
IMPERIAL WAR MUSEUM, LONDON
Clauson Papers
Dawnay papers
NATIONAL ARCHIVE, KEW (NA)
NATIONAL ARCHIVES OF CANADA
THE US NATIONAL ARCHIVES AND RECORDS
   ADMINISTRATION

## QUOTED SOURCES
(in order of appearance)

## INTRODUCTION
p4: Winston Churchill: *The World Crisis Vol 1 (1923)*

## CHAPTER 1
p10: A.W. Ewing: *The Life of Sir Alfred Ewing* (1939)
p11: Hall, Papers 3/1 (CA) or William James: *The Eyes of the Navy –
   A Biographical Study of Admiral Sir Reginald Hall* (1955)
p12: Clarke Papers 3 (CA)
p13: J.H. Burton (eds): *The Life and Letters of Walter H. Page Vol 3*
   (1922-1925)
p14: David Ramsey: *'Blinker Hall' Spymaster – The Man who Brought
   America into World War 1* (2009)
   Hall, Papers 3/1 (CA) or D. Ramsey

W. James

Sir Guy Gaunt: *The Yield of the Years – A Story of Adventure Afloat and Ashore* (1940)

p15: David Stafford: *Churchill & Secret Service* (1997)

p17: Denniston Papers 1/2 (CA) or R. Denniston: *Thirty Secret years – A.G. Denniston's Work in Signals Intelligence 1914-1944* (2007)

p18: Denniston Papers 1/3 (CA) or R. Denniston

p19: W. Churchill


## CHAPTER 2

p21: Robert K. Massie: *Castles of Steel – Britain, Germany and the Winning of the Great War at Sea* (2008)

W. Churchill

p22: W. Churchill

p25: R. Denniston

p26: Hall Papers 3/1 (CA)

p27: Penelope Fitzgerald: *The Knox Brothers* (2002)

p29: W. Churchill

p30: R. Denniston

Clarke Papers 3 (CA) or Patrick Beesley: *Room 40 – British Naval Intelligence 1914-1918* (1984)


## CHAPTER 3

p35: Mark Ellis: 'German-Americans in World War 1' from *Enemy Images in American History*, eds. Ragnhild Fiebig-von Hase, Ursula Lehmkuhl (1997) Historical money value calculation made on www.usinflationcalculator. com

Captain Henry Landau, *The Enemy Within* (1937). This was the account of the money given after the war by Germany's US paymaster, Dr. Heinrich Albert

For the full proclamation, see 'President Wilson Proclaims Our Strict Neutrality', *New York Times*, 5 August 1914

pp35–6: Ron Chernow: *The House of Morgan* (1990)

Ibid. Indeed, when Morgan learned that German investors planned to purchase Bethlehem Steel, they marshaled voting shares to block it. The grateful British exempted the House of Morgan from mail censorship, and allowed them to use an in-house code for transatlantic communication

p36: Population numbers come from www.census.gov/population/
estimates/nation/popclockest.txt
www.loc.gov/rr/european/imde/germchro.html

pp37–8: *New York Times*, 1 May 1915
'Why *Lusitania* Plans Show Gun Outlines', *New York Times*, 19 June
1915

pp39–40: P. Gannon

p44: *The Daily Chronicle*, 8 May 1915
*Tacoma Times*, 7 May 1915
*El Paso Herald*, 8 May 1915
Ibid

p45: Ibid
'No Need to Fight if Right', *New York Times*, 11 May 1915
'"I'm not here!" cries Count Bernstorff', *New York Times*, 9 May 1915

pp45–6: James W. Gerard: *My Four Years in Germany* (1917) p 173
Ibid


## CHAPTER 4

pp47–50: Malcolm Hay: *Wounded and Taken Prisoner – by an Exchanged
Prisoner* (1916)

p51: Alice Ivy Hay: *Valiant for Truth – Malcolm Hay of Seaton* (1971)

p52: Malcolm Hay: *Notes on Cryptography* in A.I. Hay

p54: R. Wilson (eds): *Frances Partridge – Diaries 1939–1972* (2001)

p55: B. Strachey: *Remarkable Relations – The Story of the Pearsall Smith
Family* (1980)

p56: B. Caine: *Bombay to Bloomsbury – A Biography of the Strachey Family*
(2005)


## CHAPTER 5

p61: Henry Landau

p63: Population figures for New York City come from *1915 Almanac and
book of Facts*, (1914) London comes in second, though a footnote
indicates 'metropolitan London' is the world's largest city with 7.2
million people. Another note indicates China is left out altogether,
as their figures are 'untrustworthy'

pp63–4: Walter Laidlaw: 'Rate of New York City's Growth', *New York
Times*, 26 June 1915
'Water Frontage Around New York', *New York Times*, 3 April, 1910

Ric Burns and James Sanders, with Lisa Ades: *New York: An Illustrated History* (2003)

p64: For a wonderfully entertaining account of von der Goltz's activities in Paris, and the international fallout, see both Captain Horst von der Goltz: *My Adventures as a German Secret Service Agent* (1918) and Barbara Tuchman: *The Zimmerman Telegram* (1985)

Count (Johann von) Bernstorff: *My Three Years in America,* (1920) and Captain Horst von der Goltz: *My Adventures as a Secret Agent* (1918)

p65–6: Von der Goltz

Chad Millman: *The Detonators* (2006)

p66: Henry Landau

pp67–8: For a full account of the hapless Horn's adventure, see French Strother: *Fighting Germany's Spies* (1918). Horn was sentenced to 18 months in a federal penitentiary in Atlanta for transporting explosives, then extradited to Canada in 1919 and sentenced to another 10 years. He was judged insane in 1921 and deported to Germany

Von Bernstorff

pp67–70: Henry Landau

Richard Spence: *Secret Agent 666* (2008)

'Keeping Posted: The Voskas', *Saturday Evening Post*, 4 May, 1940

Thomas A. Reppetto: *Battle Ground New York* (2012)

**CHAPTER 6**

p71: Franz von Rintelen: *The Dark Invader* (1936)

Nigel West: *Historical Dictionary of World War 1 Intelligence.* It states von Rintelen was born in Frankfurt an der Oder in August 1878

p72: Henry Landau

pp74–7: Von Rintelen

p76: James D. Livingston: *Arsenic and Clam Chowder: Murder in Gilded Age New York* (2010) Scheele's prominence as a European-educated scientist saw him appear as an expert witness in New York criminal trials as early as 1896

H. Landau

p76: Inspector Thomas J. Tunney and Paul Merrick Hollister: *Throttled! The Detection of the German and Anarchist Bomb Plotters* (1919)

Von Rintelen

p77: Cigar bomb number comes from H. Landau

p78: For Gaunt's own take on the war (to be taken with a grain of salt) see his *The Yield of the Years: A Story of Adventure Afloat and Ashore* (1940)
Von Rintelen

pp79–80: R. Spence
*New York Times*, 7 July 1915

pp80–2: Von Rintelen

## CHAPTER 7

p83: 'Man Who Revealed German Plan in First War Leaves Secret Service', *Milwaukee Journal*, 20 July 1942

p84: Richard Spence makes the argument that the British were behind it all, with the help of none other than Aleister Crowley
Albert Dawson was a brave and daring filmmaker, who shot some of the war's most compelling footage. When the US joined the Allied cause, he was commissioned a captain in the US Signal Corps in charge of its military photographic laboratory. See 'Shooting the Great War: Albert Dawson and the American Correspondent Film Company, 1914–1918' in Ron van Dopperen and Cooper C. Graham.

p85: Von Bernstorff

p90: T.J. Tunney

p92: Ibid

pp94–7: Keith Jeffrey: *The History of the Secret Intelligence Service 1909–1949* (2011)
Richard Spence: 'Englishmen in New York: The SIS American Station, 1915–21' in *Intelligence and National Security, 19:3.*
Joseph Pulitzer: *Reminiscences of a Secretary, Alleyne Ireland* (1914)
'Bringing Together English Speaking Peoples' in *English Speaking World* September 1919, p15.
'Norman Thwaites Wounded', *New York Times*, 10 November, 1914
Christopher Andrew: *The Defence of the Realm: the Authorized History of MI5* (2010)

## CHAPTER 8

p100: R.J. Wyatt: *Death from the Skies – The Zeppelin Raids over Norfolk – 19 Jan 1915* (1990)
C. Cover: *Zeppelins over the Eastern Counties* (2007)

p101: R. Marben: *Zeppelin Adventures* (1931)

p102: Ibid

C. Cover

J. Ferris: 'Airbandit C31 and Strategic Air Defence during the First Battle of Britain 1915–1918' in M. Dockerill and D. French (eds): *Strategy and Intelligence – British Policy during the First World War* (1996)

Marben

p103: Hugh Cleland Hoy: *40 OB or How the War Was Won* (1932)

p104: Ibid

C. Cover

H.C. Hoy

p105: J. Ferris in Dockerill + French

Captain E. Lehmann and H. Mingos: *The Zeppelins – The Development of the Airship with the Story of the Zeppelin Air raids in the World War* (1927)

p107: H. C. Hoy

Von B. Brandenfels: *Zeppelins over England* (1931)

p108: Lyn Macdonald: *Somme* (1993)

R. Marben

P.J.C Smith: *Zeppelins over Lancashire – the Story of the Air Raids on the County of Lancashire in 1916 and 1918* (1991)


**CHAPTER 9**

p113: Hall Papers 3/1 (CA) or D. Ramsey

Hall Papers 3/2 (CA)

Ibid

p114: Ibid

p115: Hall Papers 3/5 (CA) or D. Ramsey

Ibid

Hall Papers 3/5 (CA) or Martin Gilbert: *The Challenge of War – Winston S. Churchill 1914-1916* (1971)

Ibid

p116: Lord Fisher (eds A.J. Marden): *Fear God and Dreadnaught – The Correspondence of Admiral of the Fleet Lord Fisher of Kilverstone Vol 2* (1956)

W. Churchill

Hall Papers 3/5 (CA) or D. Ramsey

Hall 3/7 or D. Ramsey

p117: Lord Fisher Vol 3

Ibid
p118: Hall Papers 3/5 (CA) or P. Beesley
   Ibid
p119: Ibid
   Lord Fisher, vol 3
   Ibid
   H. C. Hoy
   Ibid
   Hall Papers 2/1 (CA)
p120: Ibid
   Hall Papers 3/4 (CA) or W. James
p121: Basil Thomson: *The Scene Changes* (1937)
p122: Ibid
   H. C. Hoy
p123: Ibid
p124: Christopher Andrew: *Secret Service – The Making of the British Intelligence Community* (1986)
   Alan Judd: *The Quest for C – Mansfield Cumming and the Founding of the Secret Service* (2000)
   Ibid
pp125–7: Compton Mackenzie: *My Life and Times – Octave Five 1915-1923* (1966)
p128: Hall Papers 3/1 (CA)
   B. Thomson

## CHAPTER 10
p130: R.L. Green: *A.E.W. Mason* (1952)
p131: Ibid
   Hall Papers 2/1 (CA)
   A.E.W Mason: *The Summons* (1920)
p132: Ibid
   R.L. Green
p133: H.C. Hoy
   A.E.W. Mason
p134: D. Ramsey
p136: P. Beesley
p137: Malcolm Hay in A.I. Hay
p138: Ibid

p139: HW3/185 (NA) or Paul Gannon: *Inside Room 40 – The Codebreakers of World War 1* (2010)

Malcolm Hay in A.I. Hay

p140: HW3/184 (NA) or P. Gannon

## CHAPTER 11

p142: Filip Nerad: *'The Irish Brigade in Germany 1914–18,'* Prague Papers on the History International Relations 2006

Colm Tóibín: *'A Whale of a Time'* in London Review of Books, October 1997

p143: F. Nerad

pp144–5 R. Spence

F. Nerad

*The First World War: Part 8: Revolution*, Channel 4 Films

The commonly cited number of Irish who served for Britain in the first World War is 200,000

p146: Geoffrey Sloan: 'The British State and the Irish Rebellion of 1916: An Intelligence Failure or a Failure of Response?' in *Journal of Strategic Security, Volume 6*

Spence also says that Adler claims the British tried to get him to kill Casement, something which they would, of course, deny

p147: G. Sloan

pp148–51: Ibid

Basil Thomson: *Queer People* (1922)

Ibid

pp151–3 'Give a Dog a Bad Name: The Curious Case of F.E. Smith and the Black Diaries of Rogers Casement', *History Today*, September 1984

p153: Brian Lewis: 'The Queer Life and Afterlife of Roger Casement', in *Journal of Sexuality, Volume 14, Number 4*

## CHAPTER 12

p155: W. James

p156: Ibid

p158: R. K. Massie

Admiral Scheer: *Germany's High Seas Fleet in the World War* (1920)

V.E. Tarrant: *Jutland – The German Perspective* (1996)

p159: N. Steele and P. Hart: *Jutland 1916* (2004)

p160: R.K. Massie

V.E. Tarrant
p161: Ad. Scheer
p162: Major R.E. Priestley: *The Signal Services in the European War of 1914–1918* (1921)
Brigadier General John Charteris: *At G.H.Q* (1931)
p163: Ibid
p164: J. Ferris (eds): *The British Army and Signals Intelligence during the First World War* (1992)
p165: F. Tuohy: *The Crater of Mars* (1929)
p166: Ibid.
F. Tuohy: *The Battle of the Brains* (1930)
p167: Ibid
p168: F. Tuohy in *The Crater of Mars*
Ibid
F. Tuohy: *The Secret Corps – a Tale of Intelligence on all Fronts* (1920)
p169: W. Langford (eds): *Somme Intelligence* (2013)
Ibid
F. Tuohy in *The Crater of Mars*

## CHAPTER 13
p171–3: Von Rintelen
p174: Robert Koenig: *The Fourth Horseman: One Man's Mission to Wage the Great War in America* (2006)
p177–80: C. Millman *The Detonators* – A 5.7 Richter scale earthquake in Agadir, Morrocco in 1960 killed 12,000 people
'Millions of Persons Heard and Felt Shock', *New York Times*, 31 July 1916
Jules Witcover: *Sabotage at Black Tom: Imperial Germany's Secret War in America, 1914–1917* (1989)
Millions of Persons Heard and Felt Shock', *New York Times*, 31 July 1916
'How Eyewitnesses Survived Explosion', *New York Times*, 31 July 1916
pp181–84: 'Held as Plotters in Black Tom Fire', *New York Times*, 10 August 1916
Ibid
H. Landau
Von Rintelen
H. Landau
Richard B. Spence: 'Englishmen in New York the SIS American

Station, 1915–21' in *Intelligence and National Security* 19:3:2004
H. Landau

## CHAPTER 14

p185–6: P. Gannon
Joachim Von Zur Gathen: 'Zimmermann Telegram: The Original
Draft' in *Cryptologia 31:2–37*
Michael S. Neiberg: 'The Zimmermann Telegram and American
Entry into World War ' www.gilderlehrman.org/history-by-era/
world-war-i/essays/zimmermann-telegram-and-american-entry-
world-war-i
p187: Janus Gerard. *My Four Years in Germany* (1917)
'War Summary', *Globe and Mail*, 10 March 1917
p189: 'Saw Villa Slay Husband', *New York Times*, 15 March 1916
pp189–92: Haldeen Braddy: *Cock of the Walk, Qui-qui-ri-qui!: The Legend
of Pancho Villa* (1955)
Mitchell Yockelson: 'The United States Armed Forces and the Mexican
Punitive Expedition Parts 1 and 2' in *Prologue Magazine* Fall 1997, Vol.
29, No. 3
www.archives.gov/publications/prologue
pp190–1: 'Hanging on Villa's Tail', *New York Times*, 28 March 1916
'All Going Well, Pershing Tells Times, But End Not in Sight' *New York
Times*, 26 March 1916
p192: Frank J. Rafalko (eds) National Counter Intelligence Centre:
*A Counter-Intelligence Reader: American Revolution to World War II,
Volume 1*
pp191–4: Von Zur Gathen
cosec.bit.uni-bonn.de/fileadmin/user_upload/publications/pubs/
gat07a.pdf
P. Gannon
For a very full account, see David Kahn: *The Codebreakers* (1996)
p195–6: B. Tuchman, p160
C. Andrew: *The Defence of the Realm*
Ibid
pp196–7: Robert Lansing: *War Memoirs of Robert Lansing* (1935)
http://babel.hathitrust.org/cgi/pt?id=mdp.39015001571010;view=1up;
seq=236
*New York Times*, 1 March 1917

p197–8: *City Germans Doubt Note is Authentic'* New York Times March 2, 1917.
Ibid
www.firstworldwar.com/source/zimmermann_speech.htm
pp197–9: C. Andrew – *The Defence of the Realm*
Ibid
B. Tuchman
*New York Times*, 19 March 1917
Ibid

**CHAPTER 15**
p203: D. Ramsey
W. James
p204: F. Katz: *The Secret War in Mexico – Europe, the United States and the Mexican Revolution* (1987)
R. L. Green
pp205–7: Ibid
p209: P. Beesley
p210: W. James
p211: D. Ramsey
p212: P. Beesley

**CHAPTER 16**
p213: Captain Parker Hitt: *Manual for the Solution of Military Ciphers* (1916)
David Kahn: *The Reader of Gentlemen's Mail: Herbert O. Yardley and the Birth of American Codebreaking* (2004)
pp214–5: John Patrick Finnegan: *The Military Intelligence Story* (1998)
Richard B. Spence: 'Englishmen in New York The SIS American Station, 1915-21' in *Intelligence and National Security 19:3* (2004)
Ibid
pp216–8: Herbert O. Yardley: *The American Black Chamber* (1931)
Ibid
Ibid
pp219–21: P.G. Wodehouse: 'The Military Invasion of America: A Remarkable Tale of the German-Japanese Invasion in 1916' in *Vanity Fair*, July 1915
Christopher Andrew: *For the President's Eyes Only* (1996)

David M. Kennedy: *Over Here: The First World War and American Society* (2004)

Thomas J. Knock: *To End All Wars* (1992)

pp221–8: *Throttled!*

H. Landau

R. Spence: *Englishmen in New York*

T.J. Tunney

Ibid

H. Landau

*New York Times*, 23 April 1918

*New York Times*, 1 May 1918

## CHAPTER 17

p227: C. Andrew

P. Beesley

p229: HW7/1 (NA) or P. Gannon

p229–30: Clarke Papers 3 (CA) or P. Beesley

p230: P. Fitzgerald

p231: P. Levy (eds): *The Letters of Lytton Strachey* (2005)

W. Churchill

R. K. Massie

p232: Alan Judd

p233: R. K. Massie

p234: ADM 137/4699 (NA) or P. Gannon

E.W.B. Gill: *War, Wireless and Wangles* (1934)

p235: Ibid

H. C. Hoy

E.W.B. Gill

p236: Ibid

p237: Ibid

p238: Hall Papers 2/1 (CA) or C. Andrew

C. Andrew

W. James

## CHAPTER 18

p241: E.W. B. Gill

p242: Ibid

p245: J. Ferris (eds)
Reginald Campbell Thompson 1876–1941 – From the proceedings of the British Academy VOL XIII (1927)
G. Clauson Letter in J. Ferris (eds)

p247: B. Westrate: *The Arab Bureau – British Policy in the Middle East 1916–1920* (1992)

p249: Y. Sheffey: *British Military Intelligence in the Palestine Campaign 1914–1918* (1998)
F. Tuohy in *The Crater of Mars*

p250: L. Von Sanders: *Five Years in Turkey* (1928)
Colonel R. Meinertzhagen: *Army Diary 1899–1926* (1962)

p251: Ibid


**CHAPTER 19**

p254: M. in A.I. Hay
P. Beesley

p255: Ibid
M. Smith and R. Erskine (eds): *Action This day* (2002)

p256: Silver Donald Cameron: 'TROTSKY IN AMHERST',
www.silverdonaldcameron.ca/trotsky-amherst
articles.philly.com/1992-04-19/news/26001794_1_shells-munitions-plant-sabotage

p257: Ibid

p260: R.E. Priestley
J. Ferris (eds)
F. Tuohy in *This is Spying*

p261: J. Ferris (eds)

p263: F. Tuohy in *This is Spying*


**CHAPTER 20**

pp265–6: 'Mid-Ocean Battle at Night', New York Times, 4 July 1917
John Keegan: *The First World War* (2000)
newsok.com/french-artist-picked-outstanding-oklahoman-as-subject-of-war-painting/article/3485815

pp267–9: Karen Kovachs: *The Life and Times of Dennis E. Nolan, 1872-1956, The Army's First G2* (1998)
J. Keegan
Nora Elizabeth Daly: *Memoirs of a World War I Nurse* (2011)

Ibid

Marc B. Powe, Edward E. Wilson: *The Evolution of American Military Intelligence* (1973)

pp270–6: H. Yardley

Ibid

David Kahn: *The Reader of Gentlemen's Mail: Herbert O. Yardley and the Birth of American Codebreaking*

H. Yardley

Ibid

David Kahn: *The Codebreakers: The Comprehensive History of Secret Communication from Ancient Times to the Internet* (1996)

Tony Crilly: *The Big Questions: Mathematics* (2011)

H. Yardley

D. Kahn: *The Codebreakers* (1973)

Douglas Porch: *The French Secret Services: A History of French Intelligence from the Dreyfus Affair to the Gulf War* (1995)

Adam Hochschild: *To End All Wars* (2012)

D. Porch

p276–8: www.texasmilitaryforcesmuseum.org/choctaw/codetalkers.htm

p278–82: Tim Cook: *Shock Troops: Canadians Fighting the Great War* (2008)

Dan Jenkins: 'Winning Trench Warfare: Battlefield Intelligence in the Canadian Corps, 1914– 1918', doctoral thesis, Carleton University (1999)

Bill Rawling: *Communications in the Canadian Corps, 1915–18 Wartime Technological Progress Revisited* in *Canadian Military History Vol. 3: Iss. 2, 1994*

T. Cook

Dan Jenkins: *The Other Side of the Hill: Combat Intelligence in the Canadian Corps, 1914–18* in *Canadian Military History: Vol. 10: Iss. 2, 2001*

## CHAPTER 21

p283: Ad. Scheer

p284: F. Toye: *For What We Have Received – An Autobiography* (1948)

p286: W. James

Denniston Papers 4/2 (CA)

Ibid

pp287–8: F. Birch and D. Knox: *Alice in ID25 – A Codebreaking Parody of Alice's Adventures in Wonderland* (1918/2007)

p289: M. Hay in A.I. Hay

p290: Ibid

P. Gannon

HW3/34 (NA) or P. Gannon

Clarke Papers 3 (CA)


**CHAPTER 22**

p293: W. James

B. Thomson

p294: C. Andrew

p295: Hall Papers 1/3 (CA)

M. Hay in A.I. Hay

p296: Ibid

p297: D. Ramsey

p298: Ibid


**CONCLUSION**

p302: Denniston Papers 1/4 (CA) or M. Batey: *Dilly – The Man who Broke Enigma* (2010)

M. Batey

Sinclair Mackay: *The Secret Lives of the Codebreakers* (2010)

p303: G. Welchman: *The Hut Six Story – Breaking the Enigma Codes* (1982)

p304: M. Batey

P. Fitzgerald

# BIBLIOGRAPHY

## BOOKS

Albert, B., *South America and the First World War – The Impact of the War on Brazil, Argentina, Peru and Chile* (1988)

Anderson, Scott, *Lawrence in Arabia – War, Deceit, Imperial Folly and the Making of the Modern Middle East* (2013)

Andrew, Christopher, *Secret Service – The Making of the British Intelligence Community* (1986)

Andrew, Christopher, *For the President's Eyes Only – Secret Intelligence and the American Presidency from Washington to Bush* (1995)

Andrew, Christopher, *The Defence of the Realm – The Authorized History of MI5* (2010)

Baker, W.J., *The History of the Marconi Company* (1970)

Barker, A.J., *The First Iraq War 1914–1918 – Britain's Mesopotamian Campaign* (2009)

Barr, J., *Setting the Desert on Fire – T.E. Lawrence and Britain's Secret War in Arabia 1916–1918* (2006)

Barthas, Louis, trans. Edward M. Strauss: *Poilu – The World War One Notebooks of Corporal Louis Barthas, Barrelmaker 1914-1918* (2014)

Batey, M., *Dilly – The Man who Broke Enigma* (2010)

Belton, James, and E.G. Odell, *Hunting the Hun* (1918)

Beach, J., *British Intelligence and the German Army 1914–1918* (PhD Thesis, 2005)

Beesley P., *Room 40 – British Naval Intelligence 1914–1918* (1984)

Berg, A. Scott, *Wilson* (2013)

Bernstorff, Count Johann von, *My Three Years in America* (1920)

Birch, F. and Knox, D., *Alice in I.D.25 – A Codebreaking Parody of Alice's Adventures in Wonderland* (1918/2007)

Blaisdell, Bob, ed, *World War One Short Stories* (2013)

Blum, Howard, *Dark Invasion – 1915 – Germany's Secret War and the Hunt for the First Terrorist Cell in America* (2014)

Brandenfels, Von B., *Zeppelins over England* (1931)

Briggs, A., *Codebreaking in Bletchley Park* (2011)

Bulow, General Von, *Experience of the German 1st Army on the Somme Battle* (1917)

Burton, J. H. (eds), *The Life and Letters of Walter H. Page*, Vol.s 1-3 (1922 & 1925)

Butler, Joseph G, *A Journey Through France in Wartime* (1917)

Caine, B, *Bombay to Bloomsbury – A Biography of the Strachey Family* (2005)

Castle, I, *London 1914–1917 – The Zeppelin Menace* (2008)

Charteris, Brig. Gen. J., *At G.H.Q* (1931)

Chernow, Ron, *The House of Morgan – An American Banking Dynasty and the Rise of Modern Finance* (2010)

Chisholm, A., *Frances Partridge – A Biography* (2009)

Churchill, W., *The World Crisis* (1923)

Cook, Tim, *Shock Troops – Canadians Fighting the Great War 1917–18* (2009)

Cover, C., *Zeppelins over the Eastern Counties* (2007)

David, S, *100 Days to Victory – How the Great War was Fought and Won* (2014)

Denniston, R, *Thirty Secret Years – A.G. Denniston's Work in Signals Intelligence 1914-1944* (2007)

Dockerill, M. and French, D. (eds), *Strategy and Intelligence – British Policy During the First World War* (1996)

Ecksteins, Modris, *Rites of Spring– The Great War and the Birth of the Modern Age* (1989)

Englund, Peter, *The Beauty and the Sorrow – An Intimate History of the First World War* (2009)

Everitt, Nicholas, *British Secret Service During the Great War* (1920)

Ewing, A.W., *The Man of Room 40 – The Life of Sir Alfred Ewing* (1939)

Ferris, J. (eds), 'Before Room 40: The British Empire and Signals Intelligence 1898–1914' from *Journal of Strategic Studies Vol. 12* (1989)

Ferris, J. (eds), *The British Army and Signals Intelligence during the First World War* (1992)

Ferris, J. (eds), 'The Road to Bletchley Park – the British Experience with Signals Intelligence 1892–1945' – from *Intelligence and National Security Vol. 17* (2002)

Fisher, Lord J. (eds Arthur J. Marden), *Fear God and Dreadnought – The Correspondence of Admiral of the Fleet Lord Fisher of Kilverstone Vol. 2 – 1904–1914 – The Years of Power* & *Vol. 3 – 1914–1920 – Restoration, Abdication, and Last Years* (1956)

Fitzgerald, P., *The Knox Brothers* (2002)

Freeman, P., 'MI1(b) and the Origins of British Diplomatic Cryptanalysis' from *Intelligence and National Security Vol. 22* (2007)

Fromkin, D., *A Peace to End All Peace – The Fall of the Ottoman Empire and the Creation of the Middle East* (2000)

Fussell, Paul, *The Great War and Modern Memory* (1975)

Gannon, P, *Inside Room 40 – the Codebreakers of World War 1* (2010)

Gaunt, Sir G, *The Yield of the Years – A Story of Adventure Afloat and Ashore* (1940)

Gerard, James W., *My Four Years in Germany* (1917)

Gilbert, Martin, *The Challenge of War – Winston S. Churchill 1914–1916* (1971)

Gilbert, Martin, *The First World War* (2002)

Gill, E.W.B., *War, Wireless and Wangles* (1934)

Goltz, Horst von der, *My Adventures as a German Secret Agent* (1918)

Gordon, A., *The Rules of the Game – Jutland and British Naval Command* (1996)

Green, R.L., *A.E.W Mason* (1952)

Gurgel, Stephen, *The War to End All Germans – Wisconsin Synod Lutherans And The First World War* (MA Thesis, 2012)

Halpern, D.G., *Naval War in the Mediterranean 1914–1918* (1987)

Hanson, N., *The First Blitz – The Secret German Plan to Raze London to the Ground in 1918* (2005)

Hart, P., *The Somme* (2006)

Hart, P., *The Great War 1914–1918* (2014)

Hay, A.I., *Valiant for Truth – Malcolm Hay of Seaton* (1971)

Hay, M., *Wounded and Taken Prisoner – by an Exchanged Prisoner* (1916)

Hazlett, A., *Electronics and Sea Power* (1975)

Hinrichs, Ernest H., *Listening In – Intercepting German Trench Communications in World War I* (1996)

Hochschild, Adam, *To End All Wars – A Story of Loyalty and Rebellion, 1914–1918*

Holmes, Richard, *Tommy – The British Soldier on the Western Front 1914–1918* (2005)

Holyrod, M., *Lytton Strachey* (1994)

Hopkirk, P., *On Secret Service East of Constantinople – The Plot to Bring Down the British Empire* (2006)

Hough, Emerson, *The Web – The Authorized History of the American Protective League* (1919)

Hoy, H.C., *40 OB or How the War Was Won* (1932)

James, L., *The Golden Warrior – The Life and Legend of Lawrence of Arabia* (1995)

James, W., *A Great Seaman – The Life of Admiral of the Fleet Sir Henry Oliver* (1951)

James, W., *The Eyes of the Navy – A Biographical Study of Admiral Sir Reginald Hall* (1955)

Jeffery, K, *MI6 – The History of the Secret Intelligence Service 1909–1949* (2011)

Jellicoe, Admiral Viscount, *The Grand Fleet 1914–1916 – Its Creation, Development and Work* (1919)

Jenkins, Dan Richard, *Winning Trench Warfare – Battlefield Intelligence in the Canadian Corps, 1914–1918* (PhD Thesis, 1999)

Jolly, W.P., *Marconi* (1972)

Jones, John Price and Hollister, Paul Merrick, *The German Secret Service in America 1914–1918* (1918)

Judd, A., *The Quest for C – Mansfield Cumming and the Founding of the Secret Service* (2000)

Kahn, David, *The Code Breakers – The Story of Secret Writing* (1996)

Kahn, David, *The Reader of Gentleman's Mail: Herbert O. Yardley and the Birth of American Codebreaking* (2004)

Katz, F., *The Secret War in Mexico – Europe, the United States and the Mexican Revolution* (1987)

Keegan, John, *The First World War* (1998)

Kennedy, David M., *Over There – The First World War and American Society* (2004)

Kennedy, P., *The Rise and Fall of British Naval Mastery* (2001)

Kennet, L., *The First Air War 1915–1918* (1991)

Knock, Thomas J., *To End All Wars – Woodrow Wilson and the Quest for a New World Order* (1992)

Koenig, Robert, *The Fourth Horseman – One Man's Mission to Wage the Great War in America* (2006)

Landau, H, *The Enemy Within – The Inside Story of German Sabotage in America* (1937)

Langford, W. (eds), *Somme Intelligence* (2013)

Lardner, James and Repetto, Thomas, *NYPD – A City and Its Police* (2000)

Lehmann, Captain E. and Mingos, H., *The Zeppelins – The Development of the Airship with the Story of the Zeppelin Air Raids in the World War* (1927)

Levy, P. (eds), *The Letters of Lytton Strachey* (2005)

Macdonald, L, *1914 – The Days of Hope* (1989)

Macdonald, L, *Somme* (1993)

Macdonald, L, *They Called it Passchendaele – The Story of the Battle of Ypres and the Men Who Fought in It* (1993)

Macdonald, L, *1915 – The Death of Innocence* (1997)

Macdonald, L, *To the Last Man – Spring 1918* (1999)

Mackay, S, *The Secret Lives of the Codebreakers* (2010)

Mackenzie, C, *My Life and Times – Octave Five 1915–1923* (1966)

Marben, R., *Zeppelin Adventures* (1931)

Marler, R. (eds), *The Letters of Vanessa Bell* (1993)

Mason, A.E.W., *The Four Corners of the World* (1917)

Mason, A.E.W., *The Summons* (1920)

Massie, R.K, *Castles of Steel – Britain, Germany and the Winning of the Great War at Sea* (2008)

Mcmeekin, S., *The Berlin–Baghdad Express – The Ottoman Empire and Germany's Bid for World Power 1898–1918* (2011)

Meinertzhagen, Colonel R., *Army Diary 1899–1926* (1962)

Menand, Louis, *The Metaphysical Club – A Story of Ideas in America* (2001)

Meyer, G.J., *A World Undone – The Story of the Great War 1914 to 1918* (2006)

Millman, Chad, *The Detonators – The Secret Plot to Destroy America and an Epic Hunt for Justice* (2006)

Mills, Bill, *The League – The True Story of Average Americans on the Hunt for World War One Spies* (2013)

Moberley, Brigadier General F.J., *The Campaign in Mesopotamia 1914–1918 4 Vols* (1923–1927)

Montefiore, H.S., *Enigma – The Battle for the Code* (2001)

Morton, James, *Spies of the First World War – Under Cover for Kings and Kaiser* (2010)

Nicholson, N. and Trautmann, Joanne (eds), *The Letters of Virginia Woolf Vols 1–6* (1970–1977)

Nicholson, V., *Among the Bohemians – Experiments in Living 1900–1939* (2003)

Occleshaw, M., *Armour Against Fate – British Military Intelligence in the First World War* (1989)

O'Collins, R (eds), *Gilbert Falkingham Clayton – An Arabian Diary* (1969)

Pincock, Stephen, *Codebreaker – The History of Codes and Ciphers, From Ancient Pharaohs to Quantum Cryptography* (2006)

Poolman, K, *Zeppelins Over England* (1960)

Popplewell, R.J., 'British Intelligence in Mesopotamia 1914–1916' from Handel M. (eds), *Intelligence and Military Operations* (1990)

Popplewell, R.J., *Intelligence and Imperial Defence – British Intelligence and the Defence of the Indian Empire 1904–1924* (1991)

Porch, D., *The French Secret Services – From the Dreyfuss Affair to the Gulf War* (1995)

Priestley, Major R.E., *The Signal Services in the European War of 1914–1918* (1921)

Proctor, Tammy M., *Female Intelligence – Women and Espionage in the First World War* (2003)

Ralph, Wayne, *Barker, VC* (1997)

Ramsey, D., *"Blinker Hall" Spymaster – The Man who Brought America into World War I* (2009)

Repetto, Thomas, *Battleground New York City – Countering Spies, Saboteurs and Terrorists Since 1861* (2012)

Rintelen, Franz von, *The Dark Invader – Wartime Reminiscences of a German Naval Intelligence Officer* (1933)

Robinson, H.R., *The Zeppelin in Combat – A History of the German Naval Airship Division 1912–1918* (1962)

Roskill, S., *Admiral of the Fleet Earl Beatty – The Last Naval Hero – An Intimate Biography* (1980)

Salvado, F.J.R., *Spain 1914–1918 – Between War and Revolution* (1999)

Sanders, L. Von, *Five Years in Turkey* (1928)

Satia, Priya, *Spies in Arabia– The Great War and the Cultural Foundations of Britain's Covert Empire in the Middle East* (2008)

Schee, Admiral, *Germany's High Seas Fleet in the World War* (1920)

Seligmann, M.S., *Spies in Uniform – British Military Intelligence on the Eve of the First World War* (2006)

Sheffey, Y., *British Military Intelligence in the Palestine Campaign 1914–1918* (1998)

Sheffey, Y., 'British Intelligence in the Middle East 1900–1918 – How Much Do We Know?' from *Intelligence and National Security Vol. 7* (1992)

Sheffey, Y., 'Institutionalised Deception and Perception Reinforcement, Allenby's Campaign in Palestine 1917–1918' from Handel, M. (ed), *Intelligence and Military Operations* (1990)

Sheldon, J., *The German Army on the Somme 1914–1916* (2005)

Sheldon, J., *The German Army at Passchendaele* (2007)

Sheldon, J., *The German Army at Vimy Ridge* (2008)

Sheldon, J., *The German Army on the Western Front 1915* (2012)

Singh S, *The Code Book – The Secret History of Codes and Code-breaking* (2000)

Skaggs, William H, *German Conspiracies in America* (1916)

Smith, M., *Station X – How the Bletchley Park Codebreakers Helped Win the War* (2011)

Smith, M. and Erskine, R. (eds), *Action This Day* (2002)

Smith, P.J.C., *Zeppelins Over Lancashire – The Story of the Air Raids on the County of Lancashire in 1916 and 1918* (1991)

Spence, Richard, *Secret Agent 666 – Aleister Crowley, British Intelligence and the Occult* (2008)

Steel, N. and Hart, P., *Jutland 1916* (2004)

Stevenson, D., *1914–1918 – The History of the First World War* (2004)

Stevenson, D., *With Our Backs to the Wall – Victory and Defeat in 1918* (2012)

Stone, N., *World War One – A Short History* (2008)

Strachan, H., *The First World War* (2006)

Strachey, B., *Remarkable Relations – The Story of the Pearsall Smith Family* (1980)

Strachey, J. and Partridge, Frances, *Lives and Letters – Julia – A Portrait of Julia Strachey* (2004)

Stafford, D., *Churchill and Secret Service* (1997)

Strother, French., *Fighting Germany's Spies* (1918)

Sykes, C., *Wassmuss* (1936)

Tarrant, V.E., *Jutland – The German Perspective* (1996)

Terraine, John, *White Heat – The New Warfare 1914–18*

Thomson, B., *Queer People* (1922)

Thomson, B., *The Scene Changes* (1937)

Townshend, C., *When God Made Hell – the British Invasion of Mesopotamia and the Creation of Iraq 1914 – 1921* (2010)

Toye, F., *For What We Have Received – An Autobiography* (1948)

Tuchman, Barbara, *The Zimmermann Telegram* (1958)

Tuchman, Barbara, *The Guns of August* (1962)

Tunney, Thomas J., *Throttled! – The Detection of the German and Anarchist Bomb Plotters in the United States* (1919)

Tuohy, F., *The Secret Corps – A Tale of Intelligence on All Fronts* (1920)

Tuohy, F., *The Crater of Mars* (1929)

Tuohy, F., *The Battle of the Brains* (1930)

Tuohy, F., *This is Spying* (1936)

Varnava, A., 'British Military Intelligence in Cyprus during the Great War' from *War in History* (2012)

Voska, E.V., *Spy and Counter Spy – The Autobiography of a Master Spy* (1941)

Weightmann, G., *Signor Marconi's Magic Box – How an Amateur Inventor Defied Scientists and Began the Radio Revolution* (2002)

Weiner, Tim, *Enemies – A History of the FBI* (2012)

Welchman, G., *The Hut Six Story – Breaking the Enigma Codes* (1982)

West, N., *Historical Dictionary of Signals Intelligence* (2012)

Westrate, B., *The Arab Bureau – British Policy in the Middle East 1916–1920* (1992)

Wilhelm II, transl. Ybarra, Thomas R., *The Kaiser's Memoirs* (1922)

Wilson R. (eds), *Frances Partridge – Diaries 1939–1972* (2000)

Witcover, Jules, *Sabotage at Black Tom – Imperial Germany's Secret War in America, 1914–1917* (1989)

Woolf, L., *An Autobiography – Vol. 1: 1880–1911 & Vol. 2: 1911–1969* (1980)

Wragg, D., *Fisher – The Admiral Who Reinvented the Royal Navy* (2009)

Wyatt, R.J., *Death from the Skies – The Zeppelin Raids over Norfolk – 19th Jan 1915* (1990)

Yardley, H., *The American Black Chamber* (1931)

## MONOGRAPHS

David George Hogarth 1862–1927 – From *Proceedings of the British Academy Vol. XIII* (1927)

Reginald Campbell Thompson 1876–1941 – From *Proceedings of the Royal Academy Vol. XXX* (1941)

US Army Intelligence and Security Command – *The Life and Times of MG Dennis E. Nolan, 1872–1956, The Army's First G2* (1998)

US Army Intelligence Center – *The Evolution of American Military Intelligence* (1973)

# INDEX

*Abhorchdienst* (German army codebreaking unit) 260

Abinger, Lord 132

*Abteilung Drei-Bai* (Section 3B) (German military intelligence unit) 35

*Abwehr* (German military and diplomatic intelligence) 303, 304

Adcock, Frank 27

Admiralty, British 3, 4, 9, 10, 11, 15, *15*, 17, 19, 21, 22, 26, 31, 38, 42, 57, 82, 93, 101, 104, 115, 117, 118, 123, 127, 128, 133, 140, 158, 160, 161, 195, 198, 212, 229, 231, 234, 284, 289, 290, 295, 299 *see also* Room 40

Admiralty, French 237

Admiralty, German 19–20

AFDGX/AFDGVX cipher 273–5

agents, field *see under individual agent name*

Air Organisation and Home Defence Against Air Raids, British 109

Aircraft 29, 102–9, 167–9, 251

Albert, Dr Heinrich 61, *61*, 63, 83–5, 87, 91

*Alert* (cable-laying ship) 4

Alfonso XIII of Spain, King 136

*Alice in ID25* (play) (Birch/Knox) 287–8

Allenby, General 250–1

*Allgemeines Funkpruchbuch* (AFB) (German code book) 106, 107

Allies *see under individual nation name*

American Correspondent Film Company 84

American Expeditionary Force (AEF) 265–72, 275–8

American Protective League (APL) 219–20

American Trench Code 276

Amiens, Battle of, 1918 279–81

Anderson, Brigadier General Francis 52

*Annie Larsen* 222, 223, 225

Arab Bureau, British 247–9

Arabia 240–1, 242, 247–9

*Arabic*, RMS 85–6

archaeology/archaeologists 28, 244, 246–7, 248, 270, 301

Archibald, James 85–6, 195

Argentina 139, 207–11

*Ariadne* 30

Armistice, 1918 286, 287

Army Cipher Disk 192, 275–6

Army War College, US 214, 215

Arnold (German agent operating in Argentina) 210–11

Asquith, Herbert Henry 114, 118

*Asturias* 43